Fundamentals of Causal Inference with R

CHAPMAN & HALL/CRC
Texts in Statistical Science Series

Recently Published Titles

Modern Data Science with R, Second Edition
Benjamin S. Baumer, Daniel T. Kaplan, and Nicholas J. Horton

Probability and Statistical Inference
From Basic Principles to Advanced Models
Miltiadis Mavrakakis and Jeremy Penzer

Bayesian Networks
With Examples in R, Second Edition
Marco Scutari and Jean-Baptiste Denis

Time Series
Modeling, Computation, and Inference, Second Edition
Raquel Prado, Marco A. R. Ferreira and Mike West

A First Course in Linear Model Theory, Second Edition
Nalini Ravishanker, Zhiyi Chi, Dipak K. Dey

Foundations of Statistics for Data Scientists
With R and Python
Alan Agresti and Maria Kateri

Fundamentals of Causal Inference
With R
Babette A. Brumback

Sampling
Design and Analysis, Third Edition
Sharon L. Lohr

Theory of Statistical Inference
Anthony Almudevar

Probability, Statistics, and Data
A Fresh Approach Using R
Darrin Speegle and Brain Claire

Bayesian Modeling and Computation in Python
Osvaldo A. Martin, Raviv Kumar and Junpeng Lao

Bayes Rules!
An Introduction to Applied Bayesian Modeling
Alicia Johnson, Miles Ott and Mine Dogucu

For more information about this series, please visit: https://www.crcpress.com/Chapman--Hall/CRC-Texts-in-Statistical-Science/book-series/CHTEXSTASCI

Fundamentals of Causal Inference with R

Babette A. Brumback

CRC Press is an imprint of the
Taylor & Francis Group, an **informa** business
A CHAPMAN & HALL BOOK

First edition published 2022

by CRC Press
6000 Broken Sound Parkway NW, Suite 300, Boca Raton, FL 33487-2742

and by CRC Press
2 Park Square, Milton Park, Abingdon, Oxon, OX14 4RN

CRC Press is an imprint of Taylor & Francis Group, LLC

ISBN: 9780367705053 (hbk)
ISBN: 9780367705091 (pbk)
ISBN: 9781003146674 (ebk)

DOI: 10.1201/9781003146674

Typeset in CMR10
by KnowledgeWorks Global Ltd.

Visit the companion website/eResources: www.routledge.com/9780367705053

To my parents

Contents

Preface

In my experience, to learn a new statistical method, it can be very helpful to reduce the problem as far as possible and work up from there. Often, I find it easiest to understand a new method in the context of a 'toy example.' In this spirit, whether learning or teaching statistical methods for causal inference, I have found much success by focusing on what I will call a 'binary dataset,' that is, a dataset in which all variables are binary. This textbook represents my effort to teach causal inference in that way. Most of the methods in the book are introduced via examples in which all variables are either naturally dichotomous or have been dichotomized by me. While I realize that this style may not appeal to everyone, I am hopeful that it will aid students in understanding the methods in this book and in learning further statistical methods that can be simplified in an analogous way. The book does not abandon more complex datasets; for example, in the exercises for Chapters 3 and 6 students learn about two-part models in the context of estimating conditional causal effects and population-averaged causal effects via standaridzation.

Research in causal inference has exploded since my first encounter with it in the late 1990s. I have chosen to focus only on fundamentals that could be covered in a one-semester course. Students are encouraged to keep learning and reaching out to the broader literature after the course ends. In keeping with a focus on fundamentals, I have striven to limit the mathematical knowledge required. I find it easiest to teach in terms of discrete probability distributions, with the understanding that students who know calculus can easily substitute integrals for sums. Chapter 2 reviews conditional probability and expectation, introduces estimation via estimating equations, and reviews sampling distributions and explains how to use the bootstrap. The mathematical level stays there for most of the remainder of the book. However, there are many hard concepts to learn, a lot of notation, and tough logical arguments throughout. Therefore, I think it is fair to say that this is not an easy textbook, despite my efforts to make it as easy as possible.

Thanks to the publisher and to the University of Florida, I had the opportunity to pilot test the book with my course on Causal Inference in Spring 2021. I reached out to undergraduates as well as graduate students, and found myself with 38 very motivated students from B.S., M.S., and Ph.D. programs in Statistics and Biostatistics and from Ph.D. programs in Education and Epidemiology. I wanted the students to have exposure to all of the exercises at the end of the chapters. Therefore, I put the students into groups of 5–6 people and assigned them to work together on all of the exercises, dividing up the work of making a presentation for each one. Then, I randomly called on students from the class to present their solutions. I made sure that the groups were relatively balanced in terms of student levels. The pilot test went well. The material is accessible to senior-level undergraduate students in Statistics, and likely also in Data Science (the program at the University of Florida was too new to have any of those). It is even more accessible to M.S. students in Biostatistics and Statistics, and not too easy to avoid challenging Ph.D. students from a variety of programs.

In my career, I have always been motivated by real examples. Therefore, I worked hard to include as many real data examples as possible in the book and in the exercises. I chose to analyze the data examples using `R` throughout the book. Some familiarity with `R` is essential prior to beginning the textbook. The appendix documents where to download `R` and which packages are necessary for the `R` code presented in the book. Having been fortunate to teach very many times out of the textbooks by J.J. Faraway that include `R` code and real data examples, I thought his approach would also work well for causal inference. Having said this, I also find it very helpful to learn and teach causal inference methods using simulated data examples, for which the causal mechanisms are known. Thus, I also make extensive use of simulations throught the text.

In very many of the examples, I do not really believe that the results can be interpreted causally. Indeed, I am generally very cautious about interpreting results causally, even when the required assumptions are more plausible. Nevertheless, many methods in this book can also be applied when assumptions do not hold, and in this case they have meaning as "adjusted associations," which are often still of interest to collaborators. Furthermore, I believe such examples still have pedagogical value.

I have taught Causal Inference three times. First, in Fall of 1999 at the University of Washington, when I was just starting out and did not know much at all; furthermore, textbooks on the topic were virtually non-existent. Second, in Spring of 2020 at the University of Florida, when I taught out of the new textbook by M.A. Hernan and J.M. Robins. This motivated me to try to present the material in my own way when I taught for the third time and piloted this textbook in Spring of 2021 at the University of Florida. I wish to thank all of my previous mentors and especially the students of those three classes, who asked many good questions that taught me a lot. Students of the third class pointed out many typos and errors in the textbook and also made suggestions for how to improve it. I am grateful to all, but wish to specifically acknowledge those who gave me written permission: Seungjun Ahn, Saurabh Bhandari, James Colee, Amy M. Crisp, Deborah Rozum, Jeremy Sanchez, Jake Shannin, Eric A. Wright, Dongyuan Wu, Xiulin Xie, Kai Yang, Xiaomeng Yuan, Wenjie Zeng, Runzhi Zhang, and Xinyi Zhang. Many thanks also to Edward Kennedy, Nandita Mitra, and Ronghui Xu for their helpful reviews.

A website for the book with datasets, `R code`, solutions to the odd exercises, and more is available by searching for the textbook at `www.routledge.com`.

1

Introduction

1.1 A Brief History

The history of causal inference is long and complex, with many thinkers from a variety of disciplines having written on the topic. A good overview from a statistician's perspective is provided by Holland (1986). Here, we briefly touch upon some prominent milestones in the field. Our earliest stop is from Aristotle (350BC), who enumerated four types of causes.

1. What a thing is made of, e.g. the porcelain of a cup
2. The form of a thing, e.g. the octave the relation of 2:1
3. "The primary source of the change or coming to rest; e.g. the man who gave advice is a cause, the father is cause of the child, and generally what makes of what is made and what causes change of what is changed."
4. "That for the sake of which a thing is done, e.g. health is the cause of walking about. ('Why is he walking about?' we say. 'To be healthy,' and, having said that, we think we have assigned the cause.)"

In statistics, causes are of Aristotle's type 3, "what causes change of what is changed." We are commonly interested in the effect of an intervention, or the cause of a health or societal problem. Type 4 causes are interesting to ponder, as these might be taken to imply that the future can cause the past, in a certain sense. However, if we translate the definition so that our intention to achieve the end is the cause, then our intention (to be healthy) can be posited to exist prior to, and thus be a type 3 cause of, the action (walking) which in turn causes (again in the sense of type 3) the end (health). If divine intervention were possible, it could be that the intention belonged to a supreme being, who then manipulated the sequence of events at least as often as necessary to reach the desired end, or fate. In that semi-deterministic or deterministic world, whether the future caused the past or the past caused the future would be irrelevant. We would be watching a movie directed by the supreme being, needlessly concerning ourselves with optimal decision-making.

Hume (1738) famously wrote in 1738 that cause and effect is not much more than the constant conjunction of events.

DOI: 10.1201/9781003146674-1

> "We have no other notion of cause and effect, but that of certain objects, which have been *always conjoined* together, and which in all past instances have been found inseparable. We cannot penetrate into the reason of the conjunction. We only observe the thing itself, and always find that, from the constant conjunction, the objects require a union in the imagination... Thus, though causation be a *philosophical* relation, as implying contiguity, succession, and constant conjunction, yet it is only so far as it is a *natural* relation, and produces a union among our ideas, that we are able to reason upon it, or draw any inference from it."

It is noteworthy that Hume did not propose any methods to verify a causal relationship other than to observe contiguity, succession, and constant conjunction. Nevertheless, shortly afterwards in 1747, James Lind undertook what is widely recognized today as one of the very first medical trials, proving a causal relationship between eating oranges and lemon and recovering from scurvy. From Brown (2005), we read that Lind took aside twelve men with advanced symptoms of scurvy "as similar as I could have them." Six pairs of men aboard the HMS *Salisbury* were thus experimented upon. The first pair were given slightly alcoholic cider. The second were given an elixir of vitriol. The third pair took vinegar. The fourth drank sea water. The fifth were fed two oranges and one lemon daily for six days, when the meager supply ran out. The sixth were given a medicinal paste and cream of tartar, which is a mild laxative. The pair who were fed the oranges and the lemons were nearly recovered after only a week. Those who had drunk the cider responded favorably, but at the end of two weeks they were still too weak to return to duty. The other four pairs did not experience good effects.

By 1846, Mill (1846), in his five canons, proposed five methods for proving cause and effect:

> "First Canon (the Method of Agreement): If two or more instances of the phenomenon under investigation have only one circumstance in common, the circumstance in which alone all the instances agree, is the cause (or effect) of the given phenomenon.
>
> Second Canon (the Method of Difference): If an instance in which the phenomenon-under investigation occurs, and an instance in which it does not occur, have every circumstance save one in common, that one occurring only in the former; the circumstance in which alone the two instances differ, is the effect, or cause, or a necessary part off the cause, off the phenomenon.
>
> Third Canon (the Joint Method of Agreement and Difference): If two or more instances in which the phenomenon occurs have only one circumstance in common, while two or more instances in which it does not occur have nothing in common save the absence of that circumstance; the circumstance in which alone the two sets of instances differ, is the effect, or cause, or a necessary part of the cause, of the phenomenon.

> Fourth Canon (the Method of Residues): Subduct from any phenomenon such part as is known by previous inductions to be the effect of certain antecedents, and the residue of the phenomenon is the effect of the remaining antecedents.
>
> Fifth Canon (the Method of Concomitant Variations): Whatever phenomenon varies in any manner whenever another phenomenon varies in some particular manner, is either a cause or an effect of that phenomenon, or is connected with it through some fact of causation."

Mill's Method of Concomitant Variation recalls Hume's constant conjunction, but Mill goes much further, and canonizes a scientific method of induction to prove causality. In the context of the scurvy trial, the first canon is satisfied because both of the patients fed lemons and oranges fully recovered. The second canon is satisfied because all of the other patients did not fully recover (although the patients given cider partially recovered, due to the small amount of vitamin C in cider). The third canon applies by virtue of the first and second canon. The fourth canon is not needed in this example, because Lind took sailors as similar as possible, and so it is not that plausible that the reason both sailors given lemon and oranges recovered is because they were somehow different from the others. The fifth canon applies to future studies of scurvy, which identified vitamin C deficiency as its consistent cause. We note that the fourth canon provides a method to adjust for *confounding* of the apparent causal effect, which we will discuss more in the next section.

In 1923, the statistician Jerzy Neyman wrote (translated by Dabrowska and Speed Neyman et al. (1990)):

> "I will now discuss the design of a field experiment involving plots... In designing this experiment, let us consider a field divided into m equal plots and let $U_1, U_2, \ldots, U_m$ be the true yields of a particular variety on each of these plots... To compare ν varieties, we will consider that many sequences of numbers, each of them having two indices (one corresponding to the variety and one corresponding to the plot): $U_{i1}, U_{i2}, \ldots, U_{im} (i = 1, 2, \ldots, \nu)$. Let us take ν urns, as many as the number of varieties to be compared, so that each variety is associated with exactly one urn. In the ith urn, let us put m balls (as many balls as plots of the field), with labels indicating the unknown potential yield of the ith variety on the respective plot, along with the label of the plot."

Thus Neyman is credited (Rubin (1990)) as being the first statistician to think in terms of *potential outcomes* (e.g. potential yields), as only one variety can be planted per plot. It is desired to compare the varieties on the same plots, but this is not possible, which is called the *Fundamental Problem of Causal Inference* by Holland (1986). The potential outcomes framework for causal inference is the subject of Chapter 3.

Another early statistical contributor to causal inference was the genetics researcher Sewall Wright, who pioneered the use of graphical models to encode causal assumptions in regression analyses (Wright (1934)). He presents the example shown in Figure 1.1, which shows the direct causes V_1 and V_2 of the effect V_0, as well as indirect causes V_4, V_5, and V_6, and the causally unrelated variable V_3. In Chapter 5, we present the modern day version of causal diagrams, called *directed acyclic graphs*, and we explain their use in adjusting for *confounding*.

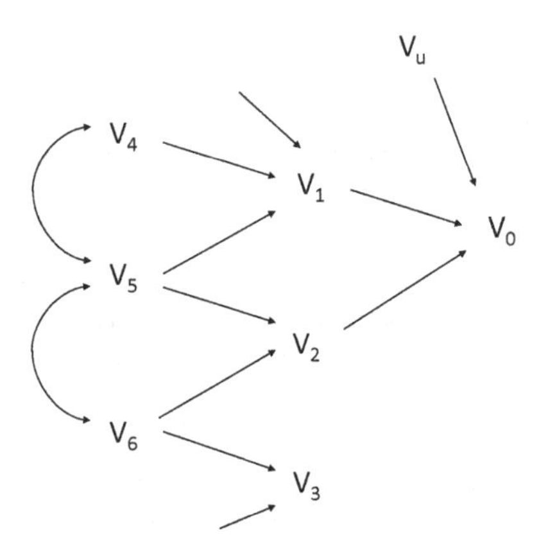

FIGURE 1.1: An Example of Wright's Path Diagrams

To study the use of streptomycin in treating tuberculosis, the Medical Research Center of Great Britain conducted the first published instance of a randomized and blind clinical trial (Hill (1990), Bhatt (2010)) allocating 107 patients to experimental and control groups using a system of random number assignments devised by Sir Austin Bradford Hill. Previously, alternating allocation, rather than randomization had been used, and Hill had been doubtful about the ethics of randomization. He justified randomization in 1947 because of a limited supply of streptomycin, and the trials were the sole means by which most Bristish patients could receive the drug. The streptomycin trial was ground breaking (Bhatt (2010)) in terms of pioneering the use of randomized treatment assignment. Today, randomized controlled trials are the gold standard for medical research, due to their ability to isolate the causal effect of the randomized treatment from that of other variables that might otherwise be associated with the treatment and the outcome.

As randomized controlled trials are not always feasible or ethical, observational studies are also heavily relied on for causal inference. In 1965, the statistician and epidemiologist Hill (1965) provided a checklist of nine items to assist in making causal judgements from observational data.

1. Strength: What is the strength of the association?
2. Consistency: Has it been repeatedly observed by different persons, in different places, circumstances and times?
3. Specificity: How specific is the association? Does the cause have many effects or is it specific to the effect under investigation?
4. Temporality: What is the temporal relationship of the association – which is the cart and which the horse?
5. Biological gradient: Is there a biological gradient, or dose-response curve?
6. Plausibility: Is the causation we suspect is biologically plausible? However, this can be a difficult criterion, because what is biologically plausible depends upon the biological knowledge of the day.
7. Coherence: The cause-and-effect interpretation of the data should not seriously conflict with the generally known facts of the natural history and biology of the disease.
8. Experiment: Occasionally it is possible to appeal to experimental, or semi-experimental, evidence. For example, because of an observed association some preventive action is taken. Does it in fact prevent?
9. Analogy: In some circumstances it would be fair to judge by analogy. For example, with the effects of thalidomide and rubella already known, we would likely accept slighter but similar evidence with another drug or another viral disease in pregnancy.

The exercises at the end of the chapter provide some examples to illustrate application of the checklist with observational studies. Much of this textbook is concerned with methods for causal inference from observational data. Chief among these are methods for adjusting for confounding, which we will begin to discuss in the next section.

During the 20th century and particularly since Hill's checklist, statisticians, social and behavioral scientists, epidemiologists, educational researchers, and econometricians have contributed much to our understanding of statistical methods for causal inference. This textbook aims to provide the fundamentals, but other textbooks will also prove helpful and offer windows into more advanced topics. The following list, in no particular order, is not exhaustive: Rothman et al. (2008), Hernan and Robins (2020), Pearl (2009), Pearl et al. (2016), Imbens and Rubin (2015), Rosenbaum (2017), Morgan and Winship (2015), and Vanderweele (2015). Additionally, Pearl and Mackenzie (2018) provide an introduction to causal inference for a popular audience, which is an excellent place to begin. Lastly, in my studies of causal inference, I have found one tenet to rest above all others, and that is that past performance is no guarantee of future results. Even if we can be reasonably sure of cause and effect in a previous study, itself an extremely tall undertaking, we have no guarantee that the relationship will hold in the future. New treatments may become available and be taken concomitantly, lifestyles may change, political upheaval may predominate, and the list of other possibilities is endless. With that caveat in mind, let us begin.

1.2 Data Examples

Throughout the text, we refer to real and hypothetical data examples to motivate and illustrate the concepts and methods. We describe some of them below.

1.2.1 Mortality Rates by Country

Confounding is a term with many definitions. A google search provides a definition from the Oxford English dictionary (August 13, 2020) of confound in standard English usage: "mix up (something) with something else so that the individual elements become difficult to distinguish." In statistics and related disciplines, a relationship between one variable and another was traditionally said to be 'confounded' when a third variable, often termed a 'lurking' variable, associated with both could explain part or all of the relationship, or could even reverse it. With the surge in popularity of causal inference, a more causal definition of confounding has become prevalent. Wikipedia's (August 13, 2020) first sentence under 'confounding' states: "In statistics, a confounder (also confounding variable, confounding factor, or lurking variable) is a variable that influences both the dependent variable and independent variable, causing a spurious association." The word 'influences' should be taken as a synonym for 'causes' in the definition. In this textbook, we introduce what we will call a *true confounder*. We modify the Wikipedia definition somewhat, to define a true confounder as a variable that influences the exposure and that also influences the outcome via a directed path that does not include the exposure (presumably, it will also influence the outcome via a directed path that includes the exposure). This definition will be further explained in Chapter 5. The existence of one or more true confounders will typically cause *confounding*; that is, the *causal effect* of the exposure on the outcome cannot be identified without either (a) adjustment involving the true confounders, or, (b) adjustment involving other variables that we will refer to simply as confounders. If there are no true confounders, then there is also no confounding. In some cases, not all of the true confounders are required for identification of the causal effect. Confounders other than true confounders are only required when a *sufficient set of true confounders* is unavailable. Econometrics uses different terminology and tends to avoid the term *confounding* in favor of the term *endogeneity*. When there are one or more true confounders, a cause is said to be *endogenous*, and otherwise, it is *exogenous*.

One of the most basic examples often used to illustrate confounding is a comparison of mortality rates by countries. For example, consider the data in Table 1.1, which presents mortality rates by age group in the US and China in 2019 (United Nations, Department of Economic and Social Affairs, Population Division (2019)) jointly with population figures from 2015 (United Nations (2020)). The overall death rate in the US in 2019 is 8.8 per 1000, whereas that in China is only 7.3 per 1000. However, we can calculate that the percent of the US population aged 65 and over is 14.6%, compared with only 9.3% in China. Looking at the age-specific mortality rates shows that the older population in the US had a mortality rate of 4.46%, whereas that in China had a rate of 5.65%. The mortality rates for the younger populations are more similar, in absolute terms: 0.268% in the US versus 0.225% in China. This is nearly an example of *Simpson's paradox* (Blyth (1972)), which refers to a relationship between two variables that reverses when a third variable is considered. That is, when we ignore age, the mortality rate is higher in the US than in China, but when we look within age category, the comparison reverses in the older age group (though not in the younger one). It is possible that if we were to break down the death rates further in the younger age group, we would find that they are uniformly lower in the US. In that case, the paradox would be complete.

TABLE 1.1
Mortality Rates by Age and Country

Country	Age Group	Deaths	Population	Rate
US	<65	756,340	282,305,227	0.002679
US	65+	2,152,660	48,262,955	0.04460
US	Overall	2,909,000	330,568,182	0.0088
China	<65	2,923,480	1,297,258,493	0.002254
China	65+	7,517,520	133,015,479	0.05652
China	Overall	10,441,000	1,430,273,973	0.0073

Interestingly, in this example, although the comparison of the US and China overall mortality rates begs to be adjusted by age, it is not obvious whether we should refer to age as a true confounder. Although age clearly 'influences' mortality, in what sense does it influence one's country? We first need to clarify what is meant, causally, by 'country' in 2019; by country, we really mean the distribution of causes of one year mortality in that country at the start of 2019. Furthermore, the distribution of age in that country clearly influences the distribution of causes of one year mortality in that country. Chapter 5 will illustrate some of the reasons to treat a variable as a confounder when it does not itself influence both exposure and outcome. Chapter 2 will explain conditional probability and expectation in terms of this example, and Chapter 6 will return to this example to illustrate *standardization*, a method we can use to compare the overall death rates had the age distributions in the two countries been identical.

1.2.2 National Center for Education Statistics

The Integrated Postsecondary Education Data System (IPEDS) is a system of interrelated surveys conducted annually by the U.S. Department of Education's National Center for Education Statistics (NCES) (`nces.ed.gov`). IPEDS annually gathers information from about 6,400 colleges, universities, and technical and vocational institutions that participate in the federal student aid programs. Table 1.2 presents a subset of the admissions data from the NCES IPEDS access database provisionally for 2018–2019. The admissions data were collected in Fall of 2018; the table includes data from institutions that enrolled at least one man and woman in Fall 2018 and that also reported SAT scores. The variable `selective` indicates that the institution admitted less than 50% of applicants. An *indicator variable*, such as `selective`, for a condition is coded as 1 to indicate the condition and 0 to indicate the absence of the condition. The variable `female` indicates that more than 60% of students admitted were women. The variable `highmathsat` indicates that the average of the 25th and 75th percentiles of the math SAT scores for enrollees was greater than 600. The number of institutions represented by a row is n.

TABLE 1.2
NCES Data

selective	female	highmathsat	n
0	0	0	435
0	0	1	87
0	1	0	420
0	1	1	37
1	0	0	50
1	0	1	104
1	1	0	55
1	1	1	29

We illustrate the concepts of *effect-measure modification* and *causal interaction* (Rothman et al. (2008)) in Chapter 4 with this example. For example, one might question whether the chance that a school admitting more than 60% women versus less than 60% is less likely to have an average math SAT score of enrollees greater than 600, and whether this chance is modified if we look within selective versus not selective schools. We might instead frame this as a question about causal interaction, and ask whether there exists a school that would have had `highmathsat` equal one if both `selective` had equaled one and `female` had equaled zero but not otherwise. In that case, selectivity and admitting fewer women would be *synergistic*. Questions of effect-measure modification and causal interaction intrinsically involve *potential outcomes* (Holland (1986)). A potential outcome to a causal event is one that we might be able to observe if the causal event happened, but would not be able to observe otherwise; Chapters 3 and 4 provide much more detail.

1.2.3 Reducing Alcohol Consumption

1.2.3.1 The What-If? Study

TABLE 1.3
The What-If? Study

T	A	H	Y	n
0	0	0	0	15
0	0	0	1	3
0	0	1	0	3
0	0	1	1	11
0	1	0	0	36
0	1	0	1	4
0	1	1	0	4
0	1	1	1	9
1	0	0	0	15
1	0	0	1	3
1	0	1	0	3
1	0	1	1	7
1	1	0	0	27
1	1	0	1	3
1	1	1	0	9
1	1	1	1	13

The WHAT-IF? (Will Having Alcohol Treatment Improve my Functioning?) study (Cook et al. (2019)) was a double-blind randomized clinical trial in which eligible women received either naltrexone 50 mg orally or placebo for 4 months, allocated in a 1:1 ratio, with assessments at baseline (i.e. at month zero, just before randomization) and at follow-up visits at 2, 4, and 7 months. Women living with HIV (WLWH) were eligible if they were 18 years or older and met past-month criteria for unhealthy alcohol use (>7 drinks/wk or >3 drinks on one single day at least twice) (National Institute on Alcohol Abuse and Alcoholism (2016)). Naltrexone is an FDA-approved medication to help reduce drinking. A subset of the data are presented in Table 1.3, where $T = 1$ denotes naltrexone and $T = 0$ placebo, $A = 1$ denotes reduced drinking (less than or equal to 7 drinks per week) for the 30 days prior to month 4, $H = 1$ denotes unsuppressed HIV viral load ($\geq$ 200 copies/ml) at baseline, and $Y = 1$

denotes unsuppressed HIV viral load at month four. The number of people represented in each row is given by n. For expository purposes, we excluded data on 7 patients missing either baseline or month 4 viral load. This choice of notation will be used throughout the book, with T and A to denote observed or randomized treatments, H to denote the medical history influencing A or T, and Y to denote the outcome.

Questions of interest that can be addressed with What-If? study data are numerous, and we have presented just a subset of the available data. A primary aim of the study was to determine if naltrexone leads to reduced drinking at 4 months. This can be addressed with a chi-squared test (with the R function `chisq.test` specifying `correct=F`) of the association between T and A (65% for naltrexone versus 62.4% for placebo reduced drinking, P = 0.72). One of the challenges with the What-If? study was that the participants were, in general, highly motivated to reduce drinking. Additionally, there was a small monetary incentive to participate, and participation required self report of heavy drinking, which meant that some participants may have consumed alcohol at a reduced rate even at baseline but reported it as heavy. These features implied limited chance of success for naltrexone to appear effective in this study population. One secondary aim was to determine if naltrexone induced improvements in clinical outcomes, such as viral load. In the naltrexone arm, 32.5% had unsuppressed viral load at month 4, versus 31.8% on placebo (P = 0.92). Had this difference been larger, but still not statistically significant, we might have wished we had planned to use H, viral load at baseline, as a *precision variable* in our analysis, to reduce the variability in our comparison. Including a precision variable in the analysis can essentially subtract some of the variability in the outcome. Incorporating precision variables is the subject of Chapter 11.

We can also investigate whether reducing drinking is associated with suppressed viral load; we observe 27.6% who reduced drinking have unsuppressed viral load at month 4, versus 40% who did not (P = 0.1). A problem with this comparison is that the apparent effect of reducing drinking might be confounded. Once again, we will use H in the analysis, but this time as a confounder rather than as a precision variable. We will also use the What-If? study data to illustrate standardization and *doubly robust estimation* (Bang and Robins (2005)) in Chapter 6. Using just the binary data of Table 1.3, we will introduce *difference-in-differences estimators* (Abadie (2005)) in Chapter 7 and compare that approach to adjusting for confounding with standardization. We will also use the binary data in Chapter 9 to illustrate *instrumental variables estimation* as another alternative method for adjusting for confounding.

1.2.3.2 The Double What-If? Study

It is not uncommon for human studies to go awry in one way or another. Humans are exceedingly complex and heterogeneous, and they vary over time in unpredictable ways. Indeed, one of the reasons to learn methods for causal inference is so that we can adjust for some of these vagaries in statistical analyses. However, it can be very helpful for learning and understanding the methods to actually *know* the causal mechanisms generating the data. For that reason, we introduce the **Double What-If? Study** (What If the What-If? Study had confirmed theory?) for illustration throughout the textbook. We simulated the data according to hyothetical known causal mechanisms depicted in Figure 1.2. We will learn about *causal directed acyclic graphs* (Pearl (1995),Greenland et al. (1999)) in Chapter 5, but an intuitive understanding of cause and effect represented via a single-headed arrow should suffice for an introduction to the study. For expository purposes, all variables are coded as binary. The variable U is latent, or unmeasured, variable representing a propensity for healthy behaviors ($U = 1$). The variables AD_0 and AD_1 represent adherence to HIV antiretroviral medication at times 0 ($AD_0 = 1$) and 1 ($AD_1 = 1$). The variables VL_0 and VL_1 represent unsuppressed HIV viral load at times 0 ($VL_0 = 1$) and 1 ($VL_1 = 1$). Note that antiretroviral adherence causes viral load even though it is coded at the same timepoint; adherence is summarizing a time period culminating at the timepoint whereas viral load is measured at that timepoint. The variable T represents randomization to naltrexone ($T = 1$) versus placebo ($T = 0$). There is no arrow into T because nothing causes it except for the randomization tool. The variable A represents reduced drinking ($A = 1$). We see that naltrexone causes reduced drinking, but so does a propensity for healthy behaviors, which also turns out to be associated with VL_1, viral load at time one (end of study) via a connection with AD_0, adherence to antiretrovirals at time zero. Throughout the book we will compare different methods for adjusting for confounding of the effect of A on VL_1, including standardization, difference-in-differences, and instrumental variables. We observe that the effect of A on VL_1 is *mediated* through AD_1. That is, the *mechanism* by which reduced drinking causes suppressed viral load is the influence of reduced drinking on adherence to antiretroviral medication, which in turn controls viral load. We will explore methods for *mediation analysis* in Chapter 12.

One of the beauties of investigating statistical methods with simulated data is that we know the truth. The R code for simulating the Double What-If? Study data is shown below. We set the randomization seed to 444 so that anyone running the code will produce the exact same data set. We simulated all binary variables using independent Bernoulli random variables with probability `prob`, where `prob` depends on the causal parents. The Bernoulli distribution models a weighted coin toss; independent Bernoulli random variables model repeated but unrelated weighted coin tosses. The Bernoulli distribution is a special case of the Binomial distribution, and therefore the `rbinom` with `size=1` generates Bernoulli random variables. We generate `n=1000` of them with probabilities given by `prob`, where the sample size of the study is `n=1000`. The variable names match the ones in Figure 1.2, except that when 'prob' is appended, it is not a random variable, but rather a collection of probabilities used for the weighted coin tosses for that variable. For example, we see that `VL0prob<-.8-.4*AD0`. This means that the probability of

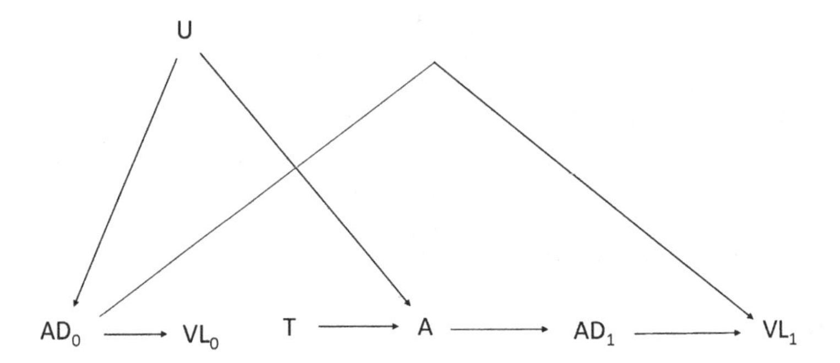

FIGURE 1.2: The Double What-If? Study

unsuppressed viral load is 0.8 without adherence to antiretrovirals at time zero, but with adherence, it is $0.8 - 0.4 = 0.4$. Therefore, patients on antiretrovirals at time zero will be more likely to have suppressed viral load, or `VL0` equal to zero. After each variable is simulated for each of the 1000 patients, they are joined together into the dataset **`doublewhatifdat`**, stored as an R data frame, convenient for our use throughout the textbook.

```
doublewhatifsim.r <- function()
{
  # Set the randomization seed for identical data upon repetition
  set.seed(444)
  # Generate 1000 independent Bernoulli random
  # variables each with probability 0.5
  U <- rbinom(n = 1000, size = 1, prob = .5)
  # Let the probability of AD0 depend on U
  AD0prob <- .2 + .6 * U
  # Generate 1000 independent Bernoulli variables
  # with varying probabilities
  AD0 <- rbinom(n = 1000, size = 1, prob = AD0prob)
  VL0prob <- .8 - .4 * AD0
  VL0 <- rbinom(n = 1000, size = 1, prob = VL0prob)
  T <- rbinom(n = 1000, size = 1, prob = .5)
  Aprob <- .05 + T * U * .8
  A <- rbinom(n = 1000, size = 1, prob = Aprob)
  AD1prob <- .1 + .8 * A
  AD1 <- rbinom(n = 1000, size = 1, prob = AD1prob)
  VL1prob <- VL0prob + .1 - .45 * AD1
  VL1 <- rbinom(n = 1000, size = 1, prob = VL1prob)
  # output the dataset
```

```
  dat <- cbind(AD0, VL0, T, A, AD1, VL1)
  doublewhatifdat <- data.frame(dat)
  doublewhatifdat
}
```

1.2.4 General Social Survey

The General Social Survey (GSS) (`www.gss.norc.org`) gathers data on contemporary United States society in order to monitor and explain trends and constants in attitudes, behaviors, and attributes. Hundreds of trends have been tracked since 1972. The data shown below contain the first 6 records of a dataset from the 2018 GSS; we have data on 2348 surveyed individuals. The GSS is conducted such a way that individuals in the US population do not have equal probability of selection into the sample. For the expository purposes of our analyses, we will assume that any resulting

```
> head(gss)
  age gt65 attend gthsedu magthsedu pagthsedu fair owngun conservative trump white female
1  43    0      1       1         0         0   NA      0            1     1     1      0
2  74    1      0       0         0         0    0      0           NA     1     1      1
3  42    0      0       1         1         0    1     NA            1     1     1      0
4  63    0      1       1         0         1    1     NA            0     0     1      1
5  71    1      1       1         0         0    1      1            1     1     0      0
6  67    1      0       1         0         0   NA      1            0     0     1      1
```

sampling bias is negligible. More advanced analyses can include the survey weights provided by the GSS to adjust for sampling bias.

We dichotomized several of the variables for our illustrative analyses. We included data on age (`age`), age greater than 65 (`gt65`), attending religious services more than once a month (`attend`), greater than a high school education (`gthsedu`), participant's mother or father having greater than a high school education (`magthsedu` or `pagthsedu`), a belief that if people got a chance they would try to be fair (rather than take advantage of you or answering it depends) (`fair`), owning a gun or revolver (`owngun`), political views that are slightly conservative or more (`conservative`), reporting casting a vote for Trump in the 2016 general election (`trump` – those who did not vote or who did not answer were coded as 0), answering 'white' to 'What race do you consider yourself' (`white`), and respondent's sex (`female`). We will use these data to illustrate adjustment for confounding using standardization in Chapter 6 and using *propensity score* methods in Chapter 10, and to illustrate mediation analysis in chapter 12.

1.2.5 A Cancer Clinical Trial

To illustrate *time-dependent confounding* (Robins (1989)) in Chapter 13, we introduce Children's Oncology Group (COG) (Pediatric Oncology Group at the time) study P9462, reported in London et al. (2010) and Brumback and London (2013). P9462 was a phase II study of relapsed neuroblastoma patients randomized to either Topotecan (TOPO, i.e. $A_1 = 0$), the standard therapy, or Topotecan + Cytoxan (TOPO/CTX, $A_1 = 1$), the experimental therapy. The primary outcome was response, originally assessed via a two-stage group sequential design. Although statistical significance was not achieved, there was a trend toward a higher response rate in those patients treated with the experimental therapy.

Although response was the primary endpoint of the P9462 study, a comparison of the proportion of patients surviving at two years ($Y = 1$) is also of interest and allows us to illustrate the analysis of *sequential treatments*, which can be subject to time-dependent confounding. A factor which complicates the survival analysis is that some patients received a bone marrow transplant (BMT) ($A_2 = 1$) after the response assessment. Nine patients in the TOPO/CTX group had a post-treatment BMT, versus only three in the TOPO group. Patients who achieved a response were more likely to have the post-treatment BMT than those who did not, in either treatment group. This raises the interesting question of how to adjust for the second treatment (the post-treatment BMT) in the comparison of survival across the first two treatments. The variable H_2, which is thought to influence both A_2 and Y, is itself influenced by A_1, and therefore meets the criteria to be a time-dependent confounder. Because A_2 is a postrandomization event, i.e. it occurs after A_1, adjusting for it in the analysis of the effect of A_1 on $Y1$ requires us to make use of the time-dependent confounder H_2 in a complex way. Two classes of statistical models which enable this are called *marginal structural models* (Robins et al. (2000)) and *structural nested mean models* (Robins (1994)). We might also be interested in knowing how to combine A_1 and A_2 to optimize survival, making use of H_2 in the decision for A_2; that is, we might wish to discover the *optimal dynamic treatment regime* (Murphy (2003)). Another wrinkle in our analyses is the problem that the trial did not produce any patients randomized to standard therapy ($A_1 = 0$) who did not have a response ($H_2 = 0$) but who had a BMT ($A_2 = 1$). Therefore, we have no data on the expected outcome Y following this sequence of events. We will suppose none of these patients were assigned $A_2 = 1$ due to an adverse expected outcome. For expository purposes, our analyses of these data will ignore the group sequential design, instead assuming an ordinary randomized clinical trial. Additionally, we will increase the sample size by a factor of 10, so that our analyses will not be limited by the small sample sizes. Finally, we will add data on 30 patients with $A_1 = 0$, $H_2 = 0$, $A_2 = 1$, and $Y = 0$, to reflect our supposition that the unobserved sequence would have led to an adverse outcome. The resulting hypothetical data are presented in Table 1.4.

As in this example, it is often the case that adjustment for time-dependent confounding is hindered by small sample sizes. In that event, such analyses are akin to *qualitative methods* (Mills and Birks (2014)). We present the analyses in this context for learning a new way to think and for planning analyses of similar studies that might be expected to produce data with the same structure.

TABLE 1.4
A Hypothetical Cancer Clinical Trial

A_1	H_2	A_2	n	n with $Y = 1$	Proportion with $Y = 1$
0	0	0	410	120	0.29
0	0	1	30	0	0.00
0	1	0	160	30	0.19
0	1	1	30	20	0.67
1	0	0	280	30	0.11
1	0	1	20	10	0.50
1	1	0	190	80	0.42
1	1	1	70	20	0.29

1.3 Exercises

1. On November 9, 2020, Pfizer reported early results from a randomized controlled trial of their coronavirus vaccine, based on

messenger RNA technology. The report stated that the vaccine appears to be at least 90% effective. To which of Hill's nine items does this evidence of cause and effect correspond?

2. At the same time, Moderna was developing a coronavirus vaccine also based on messenger RNA technology. Which of Hill's nine items suggested that the Moderna vaccine would also be effective?
3. Second generation anti-psychotics (SGAs) are helpful treatments for mental illnesses including schizophrenia and bipolar disorder. However, they are well known to cause substantial weight gain (Beasley et al. (1997)). Additionally, Lambert et al. (2006) reported on diabetes risk associated with SGAs. Which of Hill's nine items correspond to using the weight gain evidence to substantiate the cause and effect of SGAs on diabetes risk?
4. Anderson (1964) write that

   > "There seems little doubt whatever that cigarette smoking, and to a lesser extent other forms of tobacco smoking, are associated with mortality from lung cancer. This association has been amply demonstrated by both case history and cohort studies in a variety of countries, and has been thoroughly summarized at countless symposia and in innumerable review articles."

   Which of Hill's nine items does this evidence of causation address?
5. The Alpha-Tocopherol Beta Carotene Cancer Prevention Study Group (1994) reported on a randomized controlled trial of alpha-tocopherol and beta carotene in male smokers designed to validate or refute epidemiologic evidence that diets high in carotenoid-rich fruits and vegetables, as well as high serum levels of vitamin E (alpha-tocopherol) and beta carotene, reduce the risk of lung cancer. Unexpectedly, they observed a higher incidence of lung cancer among the men who received beta carotene than among those who did not. However, they found little or no effect on the incidence of cancer other than lung cancer. Which two of Hill's nine items does this new evidence of the causal effect of beta carotene on lung cancer address?
6. Lee and Skerrett (2001) found that increasing levels of physical activity are more strongly protective against all-cause mortality. Which of Hill's nine items does this evidence of causation address?
7. Boden and Fergusson (2011) concludes that a review of the literature "suggests a causal linkage between alcohol use disorders and major depression, such that increasing involvement with alcohol increases risk of depression." Which of Hill's nine items presents the greatest difficulty for concluding causation in this context?

8. Sebat et al. (2007) found that 10% of patients with sporadic autism had de novo copy number variation (CNV), versus 1% of controls, that is, individuals without sporadic autism. This is an example of a *case-control study*, in which the sampling is based on the outcome rather than the exposure. The authors conclude that these findings establish de novo germline mutation is a more significant risk factor for autism spectrum disorders than previously recognized. Which of Hill's nine items does this evidence of causation address?

2

Conditional Probability and Expectation

2.1 Conditional Probability

Table 1.1 presents the death rate in 2019 for US residents as a proportion of the population, specifically 0.0088. That proportion can also be interpreted as the probability of death in 2019 for a randomly selected US resident at the beginning of the year; i.e. the chance of death is 8.8 out of 1000, or almost 1%. Intuitively, if we knew that the selected resident were younger than 65 years, many of us would not believe such a high chance of death applied. We would focus instead on the *conditional probability* of death, given the resident's age group, or 0.2679%. In fact, even the original chance of 0.88% is a conditional probability; it is the probability of death conditional on residing in the US. This means that 0.2679% is actually the probability of death conditional both on age less than 65 and residing in the US. Often, as in this case, we can go further with our conditions and note that 0.2679% is additionally conditional on the year 2019, as opposed to 2020.

Suppose that all of our analyses will refer to the year 2019. We will bear that in mind, but not incorporate it into our statistical models. We introduce three binary random variables to model these data. Let Y be coded as 1 for being dead and 0 for being alive at the end of the year, let T be coded as 1 for residing in the US and 0 for residing in China, and let $H = 1$ denote 65+ years old and $H = 0$ denote < 65. With this notation, the conditional probability of death given age less than 65 and residing in the US is written as $P(Y = 1|T = 1, H = 0) = 0.002679$. The '$|$' symbol, or 'pipe', denotes the conditioning process. For China, we have $P(Y = 1|T = 0, H = 0) = 0.002254$. If the mortality rates for the younger population were the exactly the same in the US and China, i.e. if

$$P(Y = 1|T = 1, H = 0) = P(Y = 1|T = 0, H = 0),$$

we could state that Y is *conditionally independent* of T given $H = 0$. We would write $Y \perp\!\!\!\perp T|H = 0$. If the mortality rates for both the younger population and the older population were exactly the same in the two countries, we would write $Y \perp\!\!\!\perp T|H$. If the overall mortality rates were exactly the same in the two countries, we would write

$$Y \perp\!\!\!\perp T,$$

DOI: 10.1201/9781003146674-2

That is, Y would be *marginally independent*, or often we just say *independent*, of T. That is, the marginal probability $P(Y = 1)$ would be the same in the US and China. It is important to know and remember that conditional independence does not imply marginal independence, and vice versa. For example, if the age-specific death rates were the same in the US and China, so that $Y \amalg T|H$, it would not necessarily follow that the overall death rates are the same in the two countries, i.e. that $Y \amalg T$, because one of the countries might have proportionally more older people. Conversely, if the overall death rates were the same in the two countries, it would not follow that the age-specific death rates are the same.

We can use the *law of total probability* to recover the probability of death conditional only on residing in the US:

$$\begin{aligned} P(Y=1|T=1) = \quad & P(Y=1|T=1,H=0)P(H=0|T=1) + \\ & P(Y=1|T=1,H=1)P(H=1|T=1). \end{aligned}$$

We have that $P(Y = 1|T = 1, H = 1) = 0.04460$ from the second row of Table 1.1. The paragraph preceding Table 1.1 stated that the proportion of the US population aged 65 and over is $P(H = 1|T = 1) = 0.146$, whereas that younger than 65 is $P(H = 0|T = 1) = 1 - 0.146 = 0.854$. We can also calculate these quantities ourselves from Table 1.1; for example, $P(H = 1|T = 1) = 48,262,955/(282,305,227 + 48,262,955) = 0.146$. Using the law of total probability, then, we calculate that $P(Y = 1|T = 1) = 0.002679 * 0.854 + 0.04460 * 0.146 = 0.0088$, after rounding.

For much of this book, we will present binary datasets (that is, datasets consisting solely of binary variables) like Table 1.1 in the more convenient format of Tables 1.2 and 1.3, which present the NCES and What-If? data. From Table 1.3, we can calculate, for example, the conditional probability that viral load at 4 months is unsuppressed given a participant was assigned to placebo, did not reduce drinking, and had suppressed viral load at baseline as $P(Y = 1|T = 0, A = 0, H = 0) = 3/(3+15) = 0.1667$; likewise, the conditional probability that viral load at 4 months is unsuppressed at 4 months given a participant was assigned to naltrexone is $P(Y = 1|A = 1) = (4 + 9 + 3 + 13)/(36 + 4 + 4 + 9 + 27 + 3 + 9 + 13) = 0.276$. It is easier to calculate these probabilities in `R` if we expand the dataset to have one row per person. We repeat the first row 15 times to represent 15 people, and the second row 3 times, and so on, and we remove the column containing n. Storing these data in `whatifdat`, we can calculate $P(Y = 1|A = 1)$ via

```
mean(whatifdat$Y[whatifdat$A == 1])
```

and $P(Y = 1|T = 0, A = 0, H = 0)$ as

```
mean(whatifdat$Y[(whatifdat$T == 0) &
                 (whatifdat$A == 0) & (whatifdat$H == 0)])
```

The general structure of the law of total probability is

$$P(A|B) = \Sigma_h P(A|B, H = h)P(H = h|B),$$

where A and B are events such as $Y = 1$ and $T = 1$, and $H = h$ is itself an event. For example, for a randomly selected participant from the What-If? study population, the event $Y = 1$ occurs if viral load at 4 months is unsuppressed; the event $T = 1$ occurs if that participant is randomly assigned to naltrexone; the event $H = h$ occurs if the participant's baseline viral load is h, which could equal 0 (unsuppressed) or 1 (suppressed). The *multiplication rule* is also very useful, and can be written as

$$P(A|B, C)P(B|C) = P(A, B|C),$$

where A, B, and C are three events. If C always occurs, then we have

$$P(A|B)P(B) = P(A, B).$$

Probabilities like $P(A, B|C)$ are called *joint probabilities*, because they represent the probability that both A and B occur jointly, conditional on C. Algebraic manipulation gives us that

$$P(A|B, C) = P(B|A, C)P(A|C)/P(B|C),$$

another useful rule. The exercises give practice using these rules to make calculations using data from Tables 1.2 and 1.3.

2.2 Conditional Expectation and the Law of Total Expectation

Conditional expectation and the law of total expectation are two concepts used repeatedly in causal inference. With binary datasets, we will see that conditional expectation is tightly linked to the conditional probability. The conditional expectation quantifies what we expect to happen conditional on certain events having happened. We operationalize this using a mean; for a binary outcome coded as 0 or 1, the mean happens to be a proportion, specifically the proportion of ones. For the randomly selected resident of the US at the beginning of 2019, the conditional expectation of their mortality outcome is $E(Y|T = 1) = 0.0088$. Note that this equals $P(Y = 1|T = 1)$, computed previously. In statistical terms, conditional expectation is defined as

$$E(Y|T) = \Sigma_y y P(Y = y|T), \tag{2.1}$$

where y ranges over all of the possible values of Y. For binary Y, this reduces to $P(Y = 1|T)$. The set of conditioning variables can be larger:

$$E(Y|T, H) = \Sigma_y y P(Y = y|T, H).$$

For continuous Y, the sum is replaced with an integral and $P(Y = y|T)$ is a probability density rather than a probability mass function. We note that conditional expectation is a *linear operator*. That is,

$$E(a(T)Y_1 + b(T)Y_2|T) = a(T)E(Y_1|T) + b(T)E(Y_2|T),$$

where $a(\cdot)$ and $b(\cdot)$ are any functions of T, such as $a(T) = \alpha_0 + \alpha_1 T + \alpha_2 T^2$; note that the functions can contain any constants, such as α_0.

Analogous to the law of total probability is the *law of total expectation*, also called the *double expectation theorem*:

$$E(Y|T) = E_{H|T}\left(E(Y|H,T)\right) = \\ \Sigma_h \left\{\Sigma_y y P(Y = y|H = h, T)\right\} P(H = h|T). \tag{2.2}$$

For binary datasets, this law reduces to the law of total probability. To see this, observe that the quantity inside the curly braces reduces to $P(Y = 1|H = h, T)$, so that the right-hand side equals $P(Y = 1|H = 0, T)P(H = 0|T) + P(Y = 1|H = 1, T)P(H = 1|T)$. In the middle expression, the outer expectation is given the subscript $H|T$ to denote that we are taking the expectation of the inner expectation (which is a function of the random variables H and T) with respect to the conditional probability of H given T. We note that $E(H|T)$ can also be expressed as $E_{H|T}(H)$. We also note that the laws of total expectation and total probability apply without conditioning on T. By pretending that T is constant for everyone, i.e. that it is non-random, probabilities and expectations conditional on T are the same as they would be unconditional on T. In that case, the definition of conditional expectation at (2.1) would reduce to $E(Y) = \Sigma_y y P(Y = y)$, and the double expectation theorem would reduce similarly to $E(Y) = E_H(E(Y|H))$. Again, for continuous random variables, the sums and conditional probabilities can be replaced with integrals and conditional probability densities.

Two other useful concepts are *mean independence* and *conditional mean independence*. The random variable Y is *mean independent* of T if

$$E(Y|T) = E(Y),$$

and it is conditionally mean independent of T given H if

$$E(Y|H,T) = E(Y|H).$$

For binary datasets, mean independence and conditional mean independence are identical to the independence and conditional independence concepts that were defined previously. Concepts closely related to mean independence and conditional mean independence are *conditional uncorrelation* and *uncorrelation*. Y is conditionally uncorrelated with T given H if

$$E(YT|H) = E(Y|H)E(T|H).$$

Y is uncorrelated with T if

$$E(YT) = E(Y)E(T);$$

this is, a special case of conditional uncorrelation in the case that H is constant (and hence is unnecessary). It happens that conditional mean independence implies conditional uncorrelation, but not the other way around. To prove the forward part of this assertion, we assume $E(Y|T,H) = E(Y|H)$. Then we use the double expectation theorem to write $E(YT|H) = E_{T|H}(E(YT|T,H))$. Next we observe that $E(YT|T,H) = TE(Y|T,H)$, because expectation is a linear operator. Next, we have that $TE(Y|T,H) = TE(Y|H)$ by conditional mean independence. Finally, $E_{T|H}(T(E(Y|H)) = E(T|H)E(Y|H)$, again because expectation is a linear operator, and this proves our forward assertion. That conditional uncorrelation does not imply conditional mean independence is left as an exercise.

A statistical model for a conditional expectation is called a *regression model*. For much of our textbook, with our focus on binary datasets, we have the luxury of working with *saturated* or *nonparametric* regression models. The regression model is saturated if it does not make any assumptions beyond the basic sampling assumptions that relate the data to the underlying probabilities. For example, for a binary dataset, the model

$$E(Y|H,T) = \beta_0 + \beta_1 H + \beta_2 T + \beta_3 T * H$$

is saturated because it relates the four proportions represented by $E(Y|H,T)$ to the four parameters $\beta_0, \ldots, \beta_3$ without any constraints on the proportions. We have that $E(Y|H=0,T=0) = \beta_0$, $E(Y|H=1,T=0) = \beta_0 + \beta_1$, $E(Y|H=0,T=1) = \beta_0 + \beta_2$, and $E(Y|H=1,T=1) = \beta_0 + \beta_1 + \beta_2 + \beta_3$. Given any four proportions that we might observe in a binary dataset, we can find values of $\beta_0, \ldots, \beta_3$ that fit them. In contrast, the model

$$E(Y|H,T) = \beta_0 + \beta_1 H + \beta_2 T$$

is called *unsaturated* or *parametric*, because it makes a (parametric) modeling assumption, namely that there is no *statistical interaction* between H and T, i.e. the parameter $\beta_3 = 0$. How can we see that this imposes a constraint on the proportions? Simple algebra shows that

$$\begin{aligned}\beta_3 = {} & E(Y|H=1,T=1) - E(Y|H=1,T=0) - \\ & \{E(Y|H=0,T=1) - E(Y|H=0,T=0)\},\end{aligned}$$

so that $\beta_3 = 0$ imposes the constraint that

$$\begin{aligned}& E(Y|H=1,T=1) - E(Y|H=1,T=0) = \\ & E(Y|H=0,T=1) - E(Y|H=0,T=0).\end{aligned}$$

This is not ideal, because the constraint will not be satisfied by the vast majority of datasets.

This leads us to question: why should we ever make parametric modeling assumptions? The answer is pragmatic, because sometimes we have no better option. For example, when H and/or T are continuous random variables, even including the interaction β_3 does not remove all constraints on the conditional expectations of Y given various combinations of H and T. Alternatively, we might be interested in investigating $E(Y|T, H_1, \ldots, H_q)$ for a large number of binary variables T and $H_1, \ldots, H_q$. Including all of the two-way interactions, three-way interactions, and so forth is typically not possible because most datasets do not represent a large enough sample. That is, we typically do not have reliable enough information on the conditional mean of Y given a unique combination of T and $H_1, \ldots, H_q$ for q a relatively large number. For example, suppose T denotes a new treatment for high blood pressure, Y denotes blood pressure 1 month later, and $H_1, \ldots, H_q$ denote a large set of confounders that cause both T and Y, such as blood pressure history, age, gender, health insurance status, comorbities, and other medications. Even when those confounders are treated as binary variables, i.e. they are dichotomized, we might have very few or no observations on young women with good blood pressure a year ago and no health insurance but having diabetes and taking metformin. Making the parametric modeling assumptions allows us to *borrow information* from other individuals whom the model characterizes as relatively similar.

Sometimes, nonlinear parametric models are preferable to linear ones. The model is linear if the conditional expectation is a linear function of the parameters; otherwise, the model is nonlinear. A linear function $f(\cdot)$ of parameters such as β_1 and β_2 satisfies

$$f(\beta_1 X_1 + \beta_2 X_2) = X_1 f(\beta_1) + X_2 f(\beta_2).$$

Examples of nonlinear parametric models include loglinear and logistic models. We compare the three most common parametric models below, letting the *covariates* $X_1, \ldots, X_p$ be either functions of T and the $H_1, \ldots, H_q$ (such as T or H_1 or $T * H_1$ or even the intercept, 1) or defined more generally.

$$\begin{array}{rr} \text{Linear} & E(Y|X_1, \ldots, X_p) = \beta_1 X_1 + \cdots + \beta_p X_p \\ \text{Loglinear} & \log(E(Y|X_1, \ldots, X_p) = \beta_1 X_1 + \cdots + \beta_p X_p \\ \text{Logistic} & \text{logit}(E(Y|X_1, \ldots, X_p) = \beta_1 X_1 + \cdots + \beta_p X_p, \end{array}$$

where the function $\text{logit}(p) = \log(\frac{p}{1-p})$. Noting that $\exp(x)$ is the inverse of $\log(p)$, i.e. $\exp(\log(p)) = p$ and $\log(\exp(x)) = x$, and defining $\text{expit}(x) = \frac{\exp(x)}{1+\exp x}$ as the inverse of $\text{logit}(p)$, we can equivalently write the models as

$$\begin{array}{rr} \text{Linear} & E(Y|X_1, \ldots, X_p) = \beta_1 X_1 + \cdots + \beta_p X_p \\ \text{Loglinear} & E(Y|X_1, \ldots, X_p) = \exp(\beta_1 X_1 + \cdots + \beta_p X_p) \\ \text{Logistic} & E(Y|X_1, \ldots, X_p) = \text{expit}(\beta_1 X_1 + \cdots + \beta_p X_p). \end{array}$$

In this format, it is easier to see that the loglinear and logistic models are nonlinear models for the conditional expectation of Y.

The linear model is very suitable for modeling conditional expectations that can be both positive (greater than zero) or negative (less than zero), for example, changes in weight. The loglinear model is useful for modeling conditional expectations that are restricted to be positive, for example, healthcare expenditures. It would be awkward if some combinations of the covariates led to negative expected healthcare expenditures. The logistic model is designed for modeling conditional expectations that are restricted to be between zero and one, such as proportions; an example is mortality rate. However, when the model is saturated, the linear model is all we really need, because it will never exceed the boundaries of the observable data.

2.3 Estimation

In much of science, conditional probabilities and expectations are unknown and must be estimated. For the mortality data, the probability that a randomly selected resident of the US would die in 2019, $P(Y = 1|T = 1) = 0.0088$, could be viewed as known. In this example, we have data available on the entire population of interest. For the What-If? study, the probability that a participant who reduced drinking would have an unsuppressed viral load at month 4, $P(Y = 1|A = 1) = 0.276$, would typically be viewed as an estimate; we instead would write $\hat{P}(Y = 1|A = 1) = 0.276$. The distinction arises because the study only investigates 165 participants, 105 of whom reduced drinking. We would like to extrapolate this finding to a larger population, perhaps to all women in Florida living with HIV who meet the inclusion criteria of the study. We therefore make the assumption that 165 participants are a simple random sample from that larger population (that is, all subsets of 165 participants from the larger population are equally likely), in order to estimate $P(Y = 1|A = 1)$ for that population. Back to the mortality data, 0.0088 could also be viewed as an estimate, perhaps of the probability that a resident of the US would die in a given year ranging from 2010–2020. Sometimes the extrapolation a statistician makes is quite abstract: we assume that the residents of the US in 2019 are a simple random sample from a *superpopulation*, where the observed sample is representative of a larger whole. The superpopulation is a useful device for expressing uncertainty in a calculated probability or expectation. If we observe the entire population, and we do not appeal to a superpopulation, then there is no *sampling variability*, that is, the variability that occurs due to collecting data on just a random subset of the population. However, sometimes quantifying sampling variability is useful for measuring the uncertainty in our statistic that stems from the limited size of our sample or population. Because the size of the US population in 2019 is quite large, the sampling variability associated with $\hat{p} = 0.0088$ is quite small;

the standard error of this binomial proportion is

$$\sqrt{\frac{\hat{p}(1-\hat{p})}{n}} = \sqrt{\frac{(0.0088)(1-0.0088)}{330568182}} = 5.14 \times 10^{-6},$$

meaning that $P(Y = 1|T = 1)$ is 0.0088 give or take 5.14×10^{-6}. For the What-If? example, $P(Y = 1|A = 1)$ is 0.276 give or take 0.0436. The relative size of the standard error to the estimate is one way to measure the amount of data we have observed that bears directly on the question we are trying to answer: for the mortality example, the ratio of the standard error to the estimate is 0.0005, whereas for the What-If? example, it is 0.158, reflecting the much smaller size of the study and the much greater uncertainty.

More notation is helpful for expressing estimators of conditional probabilities and expectations more generally. We assume we have collected data on a simple random sample of individuals indexed by $i = 1, \ldots, n$, and we have observed the outcome Y and the covariates $X_1, \ldots, X_p$ for each of them; i.e. for individual i, we observe Y_i, $i = 1, \ldots, n$ and X_{ij}, $i = 1, \ldots, n$ and $j = 1, \ldots, p$. We let X_i denote the collection of X_{ij} for $j = 1, \ldots, p$, stored as a horizontal vector $(X_{i1}, \ldots, X_{ip})$, and we let X be analogous for $(X_1, \ldots, X_p)$. When we write Y and X instead of Y_i and X_i, we are referring to just one randomly selected individual from the population, within a context in which we do not need to refer to the entire sample. We let X_i^T denote the vertical representation of X_i, i.e. the transposed version. We assume an understanding of matrix multiplication, noting that the horizontal vector X_i is a $1 \times p$ matrix, and the vertical vector X_i^T is a $p \times 1$ matrix. Briefly, with β representing the vertical vector $(\beta_1, \ldots, \beta_p)^T$, by matrix multiplication we have that

$$X_i\beta = X_{i1}\beta_1 + \cdots + X_{ip}\beta_p;$$

we also have that

$$X_i^T(Y_i - X_i\beta)$$

is the vertical vector with elements

$$X_{ij}(Y_i - X_i\beta).$$

Furthermore, the expectation of a matrix with entries Z_{ij}, where i indexes the rows and j indexes the columns, is a matrix with entries $E(Z_{ij})$.

Let $i = 1, \ldots, n$ represent the participants in the What-If? study, where $n = 165$. We can estimate the unconditional expectation of the binary outcome Y as

$$\hat{E}(Y) = (1/n)\Sigma_{i=1}^{n} Y_i, \tag{2.3}$$

which returns the proportion of all of the participants with unsuppressed viral load at 4 months, equal to 0.32. This is a good estimator for $E(Y)$ because it is *unbiased*. That is,

$$E(\hat{E}(Y)) = E(Y).$$

To prove this, first we note that $E(Y_i) = E(Y)$ because we took a simple random sample from our population. Second, because expectation is a linear operator,

$$E((1/n)\Sigma_{i=1}^n Y_i) = (1/n)\Sigma_{i=1}^n E(Y_i).$$

Therefore,

$$E\left((1/n)\Sigma_{i=1}^n Y_i\right) = (1/n)\Sigma_{i=1}^n E(Y_i) = (1/n)\Sigma_{i=1}^n E(Y) = E(Y). \quad (2.4)$$

Letting $E(Y) = \beta$ denote the saturated model for that expectation, another way to express our estimator is via $\hat{\beta}$, which solves the following *estimating equation* for β:

$$U(\beta) = \Sigma_{i=1}^n (Y_i - \beta) = 0. \quad (2.5)$$

Simple algebra shows that $\hat{\beta} = \hat{E}(Y)$ of (2.3). Let

$$E(Y|X_1, \ldots, X_p; \beta) = E(Y|X; \beta)$$

be a parametric or nonparametric model for the conditional expectation of Y, such as

$$E(Y|X; \beta) = \text{expit}(X\beta);$$

note that we have inserted β into our notation $E(Y|X; \beta)$ to make the dependence on β explicit. We can construct a variety of estimating equations for β. For each of our three most common parametric models, the most common variant takes the form

$$U(\beta) = \Sigma_{i=1}^n X_i^T (Y_i - E(Y_i|X_i; \beta)) = 0, \quad (2.6)$$

where $E(Y_i|X_i; \beta)$ has exactly the same structure as $E(Y|X; \beta)$ but with Y_i and X_i in place of Y and X. For example, with the logistic model

$$E(Y|X; \beta) = \text{expit}(X\beta),$$

the estimating equation is

$$U(\beta) = \Sigma_{i=1}^n X_i^T (Y_i - \text{expit}(X_i\beta)) = 0. \quad (2.7)$$

This estimating equation can be solved using iterative methods, such as Newton-Raphson; in `R` we can use the `glm` function with the `family=binomial` option to solve it (Faraway (2016)).

For example, we can fit a logistic regression model

$$E(Y|A, T, H) = \text{expit}(\beta_1 + \beta_2 A + \beta_3 T + \beta_4 H)$$

to the What-If? data of Table 1.3 in `R` as follows.

```
> lmod <- glm(Y ~ A + T + H, family = binomial, data = whatifdat)
> lmod
Call:glm(formula = Y ~ A + T + H,
         family = binomial,
         data = whatifdat)

Coefficients:(Intercept)          A            T            H
                  - 1.523      - 0.565      - 0.225      2.744

Degrees of Freedom:164 Total (i.e. Null)
161 Residual
Null Deviance:207
Residual Deviance:151   AIC:159
```

The coefficients $(-1.523, -0.565, -0.225, 2.744)$ solve our estimating equation for $(\beta_1, \beta_2, \beta_3, \beta_4)$. We can use them to estimate $E(Y|A=1, H=1, T=1)$ at 0.606 via the estimator $\hat{E}_p(Y|A=1, H=1, T=1)$, as follows:

```
> xbeta <- sum(lmod$coef)
> exp(xbeta) / (1 + exp(xbeta))
[1] 0.60596
```

We can compare this estimate from our parametric model to the nonparametric estimate from the estimator $\hat{E}_{np}(Y|A=1, H=1, T=1)$ that we can compute straight from Table 1.3:

```
> mean(whatifdat$Y[(whatifdat$A == 1) &
                    (whatifdat$T == 1) & (whatifdat$H == 1)])
[1] 0.59091
```

The two estimates are fairly close.

The estimating equations at (2.5), (2.6), and (2.7) generally give good solutions, assuming either a nonparametric model or a correctly specified parametric model, in part because each is an *unbiased estimating equation* (van der Vaart (1998)). That is, $E(U(\beta)) = 0$; we are finding the β that solves $U(\beta) = 0$, and if we take the expectation of both sides of the equation, the answer is zero for each. We can show $E(U(\beta)) = 0$ using the law of total expectation, or double expectation theorem, which proves that

$$\begin{aligned} E(X^T(Y - \text{expit}(X\beta))) &= \\ E_X E(X^T(Y - \text{expit}(X\beta))|X) &= \\ E_X X^T E(Y - \text{expit}(X\beta)|X) &= 0, \end{aligned} \tag{2.8}$$

because $E(Y - \text{expit}(X\beta)|X)$ is zero due to linearity of expectation. We note that X^T can be taken outside the inner expectation because conditional expectation is a linear operator. The result at (2.8) follows. Therefore, $E(U(\beta)) = 0$, using an argument similar to (2.4). In addition to unbiasedness, another important feature of our estimating equations is that they are sums over individuals from a simple random sample; that is, we can write

$$U(\beta) = \Sigma_{i=1}^n U_i(\beta),$$

where, for example,

$$U_i(\beta) = X_i^T(Y_i - E(Y_i|X_i;\beta)).$$

Moreover, our simple random sample needs to be *large enough*; just how large can be difficult to know *a priori*, but for logistic regression with a binary outcome, a common rule of thumb originating from Peduzzi et al. (1996) is that both the numbers of individuals with $Y = 0$ and with $Y = 1$ need to be larger than ten times the number of parameters. For `whatifdat`, there are 112 individuals with $Y = 0$ and 53 with $Y = 1$ and only 4 parameters, which satisfies the rule of thumb. There are other *regularity conditions* estimating equations must satisfy to generate good solutions, but these generally hold for the ones we use in this textbook.

2.4 Sampling Distributions and the Bootstrap

For the What-If? data, our nonparametric *estimator* $\hat{E}_{np}(Y|A = 1, H = 1, T = 1)$ of our *estimand* $E(Y|A = 1, H = 1, T = 1)$ returns the *estimate* 0.591. While there are 165 people in the dataset, only 22 of them have $A = 1$, $H = 1$, and $T = 1$. As we discussed previously, our estimate is subject to sampling variability; if we were to redo the What-If? study with another simple random sample of 165 people from our population, we might obtain an estimate equal to 0.61, and maybe we would observe 20 participants instead of 22 with $A = 1$, $H = 1$, and $T = 1$. Just how much variability in our estimate of $E(Y|A = 1, H = 1, T = 1)$ should we expect? One way to answer this would be by reporting the standard error:

$$\sqrt{(0.591)(1 - 0.591)/22} = 0.105.$$

Another way would be by reporting the *sampling distribution* of our estimator. The sampling distribution is hypothetical: if we were to redo the study over and over and record all of the estimates into a histogram, that would represent the sampling distribution. Knowing the sampling distribution allows us to answer questions such as: what percent of our estimates would be larger than 0.75? The standard error alone does not provide this answer. For large enough samples, probability theory can be used to prove that estimators solving unbiased estimating equations that satisfy appropriate regularity conditions have a sampling distribution that is approximately Gaussian, or normally distributed (van der Vaart (1998)); the sampling distribution is centered at the true value of our estimand and it has a standard deviation equal to the true standard error of our estimator.

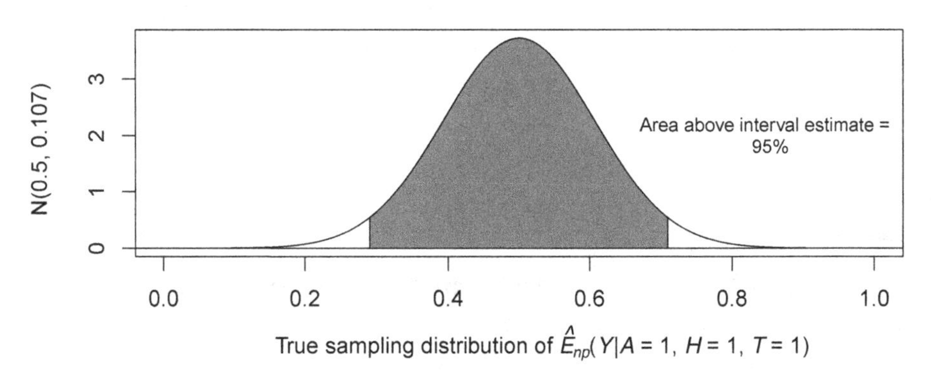

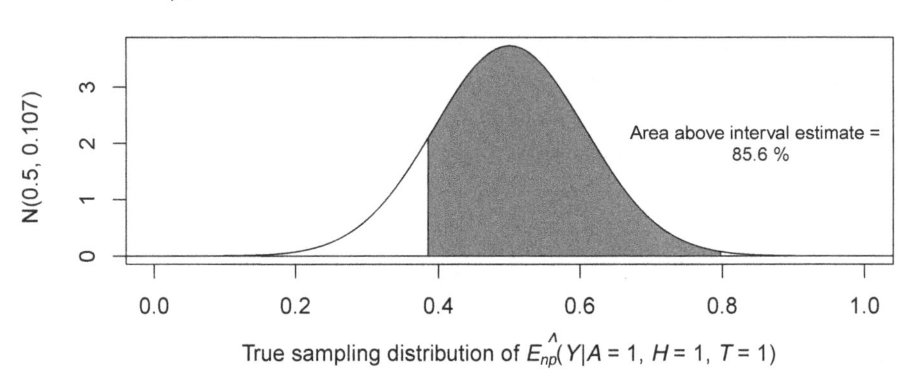

FIGURE 2.1: Percent of Estimates Contained in the 95% Probability Interval $(\hat{\mu} - 1.96\hat{\sigma}, \hat{\mu} + 1.96\hat{\sigma})$ for the Approximated Sampling Distribution $N(\hat{\mu}, \hat{\sigma})$, Where $\hat{\mu} = \hat{E}_{np}(Y|A = 1, H = 1, T = 1)$

We cannot exactly know the sampling distribution of our estimator $\hat{E}_{np}(Y|A = 1, H = 1, T = 1)$, because we do not know the true value of $E(Y|A = 1, H = 1, T = 1)$ nor the true value of its standard error. One might think to approximate the sampling distribution by a normal distribution $N(\mu, \sigma)$ with mean μ set equal to $\hat{\mu} = \hat{E}_{np}(Y|A = 1, H = 1, T = 1)$ and σ set equal to the estimated standard error $\hat{\sigma}$. Thus we would approximate the sampling distribution of our estimate as $N(0.591, 0.105)$. However, this is not really a good approximation. For example, it is not true that 95% of estimates returned by $\hat{E}_{np}(Y|A = 1, H = 1, T = 1)$ are contained within the 95% probability interval for the approximated sampling distribution, i.e. $(0.591 - 1.96 * 0.105, 0.591 + 1.96 * 0.105) = (0.385, 0.797)$; rather, that probability will typically be significantly lower than 95%. This is because 0.591 is not equal to the true value of $E(Y|A = 1, H = 1, T = 1)$. For example, suppose the true value were 0.5; then the percent of estimates contained in the 95% probability interval for the approximated sampling distribution obtained from the estimate 0.591, i.e. in (0.385, 0.797), would be significantly lower than 95%, because the interval would be significantly off center. As shown in Figure 2.1, the percentage is approximately 85.6%.

Because the majority of the estimates will lead to intervals that are signficantly off center, the chance that a random interval constructed from one sample will contain a random estimate constructed from a second sample is actually only around 83–84% when the sample size is very large and the estimator is exactly normally distributed. This can be compared to the chance that the same random interval contains the true estimand, which is very close to 95%. These chances are computed with the R function `sim.r`, below:

```
sim.r <- function ()
{
  countb <- countg <- 0
  n <- 10000
  nsim <- 10000
  for (i in 1:nsim) {
    # first sample
    # generate n independent standard normal random variables
    y <- rnorm(n)
    # compute the mean and standard error of the mean
    muhaty <- mean(y)
    sehaty <- sqrt(var(y) / n)
    # second sample independent from first
    x <- rnorm(n)
    muhatx <- mean(x)
    # check coverage of second sample mean by
    # random interval from first sample
    check1b <- (muhatx < muhaty + 1.96 * sehaty)
    check2b <- (muhatx > muhaty - 1.96 * sehaty)
    # check coverage of true mean by random
    # interval from first sample
    check1g <- (0 < muhaty + 1.96 * sehaty)
```

```
    check2g <- (0 > muhaty - 1.96 * sehaty)
    if (check1b & check2b)
      countb <- countb + 1
    if (check1g & check2g)
      countg <- countg + 1
  }
  # return the two coverages
  list(bad = countb / nsim, good = countg / nsim)
}
> sim.r()
$bad
[1] 0.8374
$good
[1] 0.948
```

The usual way to interpret the 95% *confidence interval* $(0.591 - 1.96 * 0.105, 0.591 + 1.96 * 0.105)$ is as follows: when we repeatedly redo the What-If? study, the intervals we make using this method will include the true $E(Y|A = 1, H = 1, T = 1) = \mu$ approximately 95% of the time. This is the correct interpretation taught in introductory statistics classes. When the sampling distribution of $\hat{\mu}$ is $N(\mu, \sigma)$, it follows that

$$P(-1.96 < \frac{\hat{\mu} - \mu}{\sigma} < 1.96) = 95\%;$$

multiplying through by σ, subtracting $\hat{\mu}$, and changing signs yields

$$P(\hat{\mu} - 1.96\sigma < \mu < \hat{\mu} + 1.96\sigma) = 95\%.$$

It turns out that substituting $\hat{\sigma}$ for σ results in a negligible change. It is common to say that the values for $E(Y|A = 1, H = 1, T = 1)$ contained in the confidence interval are all plausible. That is, even though we estimated $E(Y|A = 1, H = 1, T = 1)$ at 0.591, values as small as 0.385 and as large as 0.797 are still plausible. Returning to our question about the percentage of estimates that would be larger than 0.75, we actually *can* use our approximate sampling distribution to compute this relatively accurately. We find it is approximately 6.5%, computed with `1-pnorm(.75,mean=.591,sd=.105)`. In fact, we do have relatively good estimates of the mean and standard deviation of our sampling distribution, and they are good enough for some kinds of assessments, like this one, but they are not good enough for other kinds of assessments, as we saw above.

For our nonparametric estimator $\hat{E}_{np}(Y|A = 1, H = 1, T = 1)$, *large sample theory* provides us with a good approximation of its true sampling distribution, i.e. $N(\mu, \sigma)$, with $\mu = E(Y|A = 1, H = 1, T = 1)$ and $\sigma = \sqrt{\mu(1 - \mu)/n}$. Refer to Lehmann (1999) for a gentle introduction to large sample theory. As we have seen, replacing μ and σ with estimates leads to some problems, but we can still form an interpretable 95% confidence interval. In many situations, we do not have a good estimate of σ. In those cases, we often can turn to the *bootstrap* (Efron and Tibshirani (1993)), so named because estimating the unknown σ is in many ways as difficult as pulling oneself up by one's own bootstraps. Even in situations where a good estimate is theoretically possible, sometimes computing it is difficult, and once again, the bootstrap can help. We will illustrate this by computing the bootstrap confidence interval for our parametric estimator $\hat{E}_p(Y|A = 1, H = 1, T = 1)$, which returned the estimate 0.606 for the What-If? data.

There are many varieties of the bootstrap, including the nonparametric bootstrap, which we will present, and parametric versions, which we will not use in this textbook. For the goal of approximating the sampling distribution, we would ideally redo the What-If? study repeatedly and collect all of the estimates returned by $\hat{E}_p(Y|A = 1, H = 1, T = 1)$. As this is hardly feasible, instead we redo the study by resampling the individuals in our sample, with replacement, until we have another sample of the original sample size n. One such resampling is called a bootstrap sample. Suppose Bob, Judy, and Ron were the first three individuals sampled. In a bootstrap sample, we might observe Bob three times and Judy once and Ron not at all. For each bootstrap sample b, $b = 1, \ldots, B$, we reapply our estimator $\hat{E}_p(Y|A = 1, H = 1, T = 1)$ to obtain a bootstrap estimate. It is common to take $B = 500$ or 1000 bootstrap samples; therefore we wind up with that many bootstrap estimates. The sampling distributions of the estimators we obtain from the estimating equations in this textbook are all approximately normally distributed, by large sample theory. Taking linear functions of those estimators, such as summing the coefficient estimates for the parametric model for $E(Y|A = 1, H = 1, T = 1)$, does not affect the normality of the sampling distribution. We can use the standard deviation of the nonparametric bootstrap estimates, $\hat{\sigma}_{boot}$, to estimate the standard error of our original estimate, which we could denote by $\hat{\mu}$. Then we can form a 95% bootstrap confidence interval as $\hat{\mu} \pm 1.96\hat{\sigma}_{boot}$. A nuance of the nonparametric bootstrap is that we could choose to resample just the 22 individuals with $A = 1$, $H = 1$, and $T = 1$ so that only Y is random, or we could choose to resample all 165 individuals, so that A, H, and T are also random, and we would then observe different numbers (not always 22) of individuals with $A = 1$, $H = 1$, and $T = 1$ across the bootstrap samples. Both methods are correct; the first method estimates the sampling distribution conditional on $A = 1$, $H = 1$, and $T = 1$, and the second estimates the unconditional sampling distribution.

We can use the `boot` package in `R` with function `boot` to estimate a bootstrap confidence interval for $\hat{E}_p(Y|A = 1, H = 1, T = 1)$. We also need the `car` package in order for the `summary()` function to operate on `boot` objects the way we describe. We resample all 165 participants to estimate the unconditional sampling distribution as follows:

```
lmodboot.r <- function ()
{
  # estimate the parameters of the logistic model
  # and use them to compute the estimand (xbeta)
```

```
  estimator <- function(data, ids)
  {
    dat <- data[ids, ]
    coef <- glm(Y ~ A + T + H, family = binomial, data = dat)$coef
    xbeta <- sum(coef)
    xbeta
  }
  # compute bootstrap confidence intervals
  boot.out <- boot(data = whatifdat,
                   statistic = estimator,
                   R = 1000)
  # logit of the original estimate
  logitest <- summary(boot.out)$original
  # standard error of the logit of the original estimate
  SE <- summary(boot.out)$bootSE
  # confidence interval lower and upper bounds on the
  # logit scale
  logitlci <- logitest - 1.96 * SE
  logituci <- logitest + 1.96 * SE
  # exponentiate the estimate and confidence interval
  est <- exp(logitest) / (1 + exp(logitest))
  lci <- exp(logitlci) / (1 + exp(logitlci))
  uci <- exp(logituci) / (1 + exp(logituci))
  list(est = est, lci = lci, uci = uci)
}
> lmodboot.r()
$est
[1] 0.60596
$lci
[1] 0.41638
$uci
[1] 0.76823
```

Thus the bootstrap confidence interval for $E(Y|A = 1, H = 1, T = 1)$ using our parametric model is (0.416, 0.768). As the bootstrap uses a collection of random resamples, it will give slightly different results upon reuse. For example, when we ran it a second time it returned the interval (0.423, 0.763). Note that we used the bootstrap to first compute the lower and upper limits of the confidence interval on the logit scale, because $\text{logit}(E(Y|A = 1, H = 1, T = 1))$ is a linear function of the coefficients. Then we used the expit transform on those limits. The principle is if (lci,uci) is a 95% confidence interval for an estimand e, then (g(lci),g(uci)) is a 95% confidence interval for $g(e)$; in our application, $g(\cdot)$ is the expit$(\cdot)$ function. We could have computed the confidence interval directly for $E(Y|A = 1, H = 1, T = 1)$, but the large sample properties of that method are not as good; a larger sample is required for good accuracy.

Finally, we compare the results of our nonparametric estimation and parametric estimation of $E(Y|A = 1, H = 1, T = 1)$. Our nonparametric estimator and 95% confidence interval are 0.591 (0.385, 0.797), whereas for our parametric estimator we compute 0.606 (0.416, 0.768). Although the estimates are quite close, the confidence interval is narrower for the parametric estimator. This phenomenon stems from the addition of the *parametric modeling assumptions*, that is, our parametric model constrains some of the nonparametric coefficients to equal zero; the parametric model is

$$E(Y|A,H,T) = \beta_1 + \beta_2 A + \beta_3 H + \beta_4 T,$$

whereas the nonparametric model is

$$E(Y|A,H,T) = \beta_1 + \beta_2 A + \beta_3 H + \beta_4 T + \beta_5 AH + \beta_6 AT + \beta_7 HT + \beta_8 AHT.$$

Hence the parametric model assumes $\beta_5 = \beta_6 = \beta_7 = \beta_8 = 0$. These assumptions lead to a less variable estimator, i.e. enhanced *statistical efficiency*, which in turn leads to a narrower range of plausible values for the estimand. However, the price one pays is the possibility that the assumptions are wrong, which can lead to bias in the estimate and confidence interval. This is known as the *bias-variance tradeoff*.

2.5 Exercises

1. Use the data from Table 1.3 to show empirically that both sides of the equation
$$P(B,C) = P(B|C)P(C)$$
are equal, with B the event that $A = 1$ and C the event that $T = 1$.
2. Use the data from Table 1.2 to show empirically that both sides of the equation
$$P(A|B,C) = \frac{P(B|A,C)P(A|C)}{P(B|C)}$$
are equal, with A the event that `highmathsat` equals 1, B the event that `female` equals 1, and C the event that `selective` equals 1.
3. With respect to the data from Table 1.3, let A be the event that $T = 1$, B be the event that $H = 1$, and $\bar{A}$ be the event that $T = 0$. Show mathematically (using the expression for $P(A|B)$ provided in the textbook) and empirically that
$$P(A|B) = \frac{P(B|A)P(A)}{P(B|A)P(A) + P(B|\bar{A})P(\bar{A})}.$$
4. Consider two random variables X and Y such that
$$P(X = 1, Y = 1) = P(X = 2, Y = -2) = P(X = 3, Y = 1) = 1/3.$$

Use this distribution to show that uncorrelation $E(XY) = E(X)E(Y)$ does not imply mean independence $E(Y|X) = E(Y)$.

5. For the dataset `whatifdat`, compare nonparametric and parametric estimates of $E(Y|A = 1, H = 0, T = 1)$ and construct bootstrap confidence intervals for each.

3

Potential Outcomes and the Fundamental Problem of Causal Inference

Coronavirus disease 2019 (COVID-19) causes diffuse lung damage that may progress to respiratory failure and death. The RECOVERY trial was a randomized, controlled, open-label, adaptive, platform trial comparing several possible treatments for COVID-19. We pulled results from a preliminary report on June 22, 2020 (RECOVERY Collaborative Group (2020)) for the comparison of dexamethasone 6 mg given once daily for up to ten days versus usual care alone in terms of 28-day mortality. Although the report presents a more sophisticated analysis, our simple comparison yields essentially identical results. The data are presented in Table 3.1, where $T = 1$ for dexamethasone and $T = 0$ for usual care, and $Y = 0$ for death within 28 days and $Y = 1$ for survival for at least 28 days; n is the number of patients per row. We calculated $\hat{E}(Y|T = 1) = 0.823$ and $\hat{E}(Y|T = 0) = 0.754$, with a chi-squared test of the difference yielding a p-value of 0.032. These results suggest that dexamethasone can increase 28-day survival, to some extent.

TABLE 3.1
A Randomized Trial of Dexamethasone versus Usual Care

T	Y	n
0	0	1065
0	1	3256
1	0	454
1	1	2104

DOI: 10.1201/9781003146674-3

3.1 Potential Outcomes and the Consistency Assumption

We will use the potential outcomes framework for causal inference in this textbook. The RECOVERY trial raised a natural question for citizens of the world in the summer of 2020: if hospitalized for COVID-19, should we hope to receive dexamethasone? That is, would receiving it cause our survival? Thinking through this question carefully, what we really would like to do is compare our 28-day survival were we to receive dexamethasone with our 28-day survival were we to receive usual care. That is, we would like to compare our potential outcome $Y(1)$ to treatment with $T = 1$ with our potential outcome $Y(0)$ to treatment with $T = 0$. Throughout the text, we will use parenthetical notation, such as $Y(t)$, to denote potential outcomes, such as the potential outcome to $T = t$. We can contrast the potential outcomes $Y(0)$ and $Y(1)$ with the observable outcome Y that we would experience were we enrolled in the study.

The utility of the potential outcomes framework hinges on the validity of a *consistency assumption* (see Cole and Frangakis (2009)) that links potential outcomes to observed outcomes. This assumption states that if a participant is assigned $T = t$, the observed outcome Y equals, or is consistent with, the potential outcome $Y(t)$; that is, we can write $Y = Y(T)$. For a binary treatment, we can also write

$$Y = Y(0)(1 - T) + Y(1)T.$$

Under the consistency assumption, we get to observe one potential outcome per study participant. While the potential outcomes are hypothetical and postulated to exist before treatment is assigned, the observed outcome does not exist until it is assessed, which is after treatment is assigned.

A major problem for causal inference is that we can only observe one potential outcome per participant. Therefore, direct comparisons like $Y(1)$ versus $Y(0)$ are impossible. Holland (1986) called this problem the *Fundamental Problem of Causal Inference* (FPCI). Thus, we can never know for sure whether dexamethasone would help us or not. An important feature of causal inference is that there must be a basis for comparison. Suppose we only observe $Y = Y(1) = 1$. Can we then say that dexamethasone caused that outcome? Without knowing $Y(0)$, we do not know whether the patient would have died without dexamethasone; if $Y(0) = Y(1)$, then receipt of dexamethasone does not matter, and it is not causal for that patient. In the next section, we present two possible routes for circumventing the FPCI.

Recall that $Y = 0$ corresponds to death within 28 days and $Y = 1$ to 28-day survival. If we were able to know $Y(0)$ and $Y(1)$ for the patients in the study, we could classify them into four *causal types*:

$$\begin{aligned} \text{Doomed:} &\quad Y(0) = Y(1) = 0 \\ \text{Responsive:} &\quad Y(0) = 0, Y(1) = 1 \\ \text{Harmed:} &\quad Y(0) = 1, Y(1) = 0 \\ \text{Immune:} &\quad Y(0) = Y(1) = 1 \end{aligned}$$

As we only observe Y, and not the potential outcomes, we cannot pinpoint the causal type. However, if a treated patient has $Y = 1$, the patient must be responsive or immune, whereas $Y = 0$, corresponds to doomed or harmed. For an untreated patient, $Y = 1$ implies harmed or immune, whereas $Y = 0$ implies doomed or responsive. For some studies, researchers may believe that the treatment cannot be harmful. In that case, a treated patient with $Y = 0$ must be doomed, while an untreated patient with $Y = 1$ must be immune.

It is worth emphasizing that the consistency assumption requires the potential outcomes to be well-defined. For example, in vaccine trials, even with a perfect vaccine, a participant's potential outcome to not getting the vaccine could depend on what percent of the study population is vaccinated and on how much that participant interacts with individuals outside of the study population. Let $Y = 1$ indicate infection and $Y = 0$ no infection, and $T = 1$ indicate receipt of a perfect vaccine and $T = 0$ no receipt. If everyone but one participant is vaccinated, and if that person only has contacts within the study population, then that person will not be exposed to the infectious disease, so that $Y(0) = 0$. On the other hand, if only half the study population is vaccinated, then that same unvaccinated participant will have a chance of exposure to the infectious disease, so that $Y(0)$ could equal 1. This example highlights that we cannot use the potential outcomes framework unless we can carefully define the potential outcomes. We could state that $Y(t)$ pertains specifically to a trial with half of the participants randomized to the vaccine. Extrapolation to a situation in which 80% of the population adopts a vaccine is tenuous.

3.2 Circumventing the Fundamental Problem of Causal Inference

Holland (1986) presented two solutions to the fundamental problem of causal inference. The first he called the *scientific solution*, in which we could use scientific theory to measure both potential outcomes. For example, although we can only know just one of $Y_i(0)$ or $Y_i(1)$ for participant i of a study, if we believe participant j is essentially identical to participant i with respect to the two potential outcomes, we could assign $T_i = 1$ and $T_j = 0$ so that $Y_i = Y_i(1)$

and $Y_j = Y_j(0) = Y_i(0)$. Thus a contrast of the observables Y_i and Y_j would be a contrast of the unobservables $Y_i(1)$ and $Y_i(0)$. An example might be testing a new diet on genetically identical mice. Alternatively, an appeal to scientific theory might be used to impute $Y_i(0)$, the potential outcome to no treatment. For example, propofol ($T_i = 1$) is a commonly used sedative ($Y_i = 1$) for colonscopy; an anesthesiologist can safely assume that patient i would not be sedated with $T_i = 0$; that is, $Y_i(0) = 0$. After administering $T_i = 1$, we would observe $Y_i = Y_i(1)$ and thus whether propofol causes sedation. When we can assume that there are no *carry-over effects* of treatment T, we can observe individual i's outcome to $T_i = 0$ at one time point, i.e. $Y_{i1}(0)$ and to $T_i = 1$ at a second time point, i.e. $Y_{i2}(1)$, and assume either $Y_{i1}(0) = Y_{i2}(0)$ or $Y_{i2}(1) = Y_{i1}(1)$ so that a contrast such as $Y_{i2}(1) - Y_{i1}(0)$ essentially equals $Y_i(1) - Y_i(0)$. An example might be testing whether ibuprofen causes headache relief.

Holland called the second solution the *statistical solution*, in which we randomly assign treatments to individuals. For a randomized clinical trial with two treatments (or a treatment and usual care or placebo) and an assignment probability that does not depend on any other data, randomization and the consistency assumption together imply that

$$(Y(0), Y(1)) \amalg T, \tag{3.1}$$

because the potential outcomes exist prior to treatment, and the treatment is randomized independently of any existing data. The independence at (3.1) implies mean independence, that is, $E(Y(0)) = E(Y(0)|T = 0)$. Then, the consistency assumption implies that $E(Y(0)|T = 0) = E(Y|T = 0)$, so that $E(Y(0)) = E(Y|T = 0)$. Similarly, $E(Y(1)) = E(Y|T = 1)$. This means that we can estimate the expectation of the unobservables $E(Y(0))$ and $E(Y(1))$ with $\hat{E}(Y(0)) = \hat{E}(Y|T = 0)$ and $\hat{E}(Y(1)) = \hat{E}(Y|T = 1)$. Therefore, rather than contrasting the individual outcomes $Y(1)$ and $Y(0)$ for one person, we contrast the average outcomes $E(Y(1))$ and $E(Y(0))$ for the population. With the COVID-19 example, we can say that dexamethasone causes an increased chance of 28-day survival *on average* for the study population. Extrapolation of this result is fraught with difficulty. First, though the difference between 75.4% and 82.3% is statistically significant, it is still possible that it was due to chance. Second, the study population may no longer be relevant. New treatments might have been developed that were not in use at the time of study, or hospital protocols may have changed, or the population of patients may have changed; it is unknown how dexamethasone would work in conjunction with newer treatments or altered hospital protocols or a different case mix of patients. Third, and most important, extrapolating an average result to an individual is problematic. There may even be individuals in the study who were harmed by dexamethasone, but they were more than counterbalanced by individuals who were helped by it. And for the majority of patients, the outcome was the same with or without dexamethasone.

For many scientific questions that cannot be addressed using the scientific solution to the FPCI, randomized studies are not feasible, due to ethical considerations or pragmatic reasons. In those cases, we are forced to rely on observational studies, and we must try our best to adjust for confounding. As we will see in the textbook, there are several ways to do this, but a primary method relies on assuming that receipt of treatment effectively followed a *stratified randomized trial*, so that

$$(Y(0), Y(1)) \amalg A|H, \tag{3.2}$$

where $Y(a)$ is the potential outcome to $A = a$, A is completely randomized within strata defined by H, and H is the set of confounders thought to be important for adjustment. The conditional independence at (3.2) is sometimes referred to as *exchangeability* (see, for example, Greenland and Robins (2009)) or as *ignorability* (see, for example, Rosenbaum and Rubin (1983)). We also require a *positivity assumption*, that is,

$$1 > P(A = 1|H) > 0, \tag{3.3}$$

which asserts that individuals with a given level of H have a positive chance of receiving either $A = 1$ or $A = 0$. For example, suppose the obsertional study is prone to *confounding by indication* (Slone et al. (1979)), that is, patients who are indicated for the treatment, and hence are sicker, tend to receive the treatment. In this case, the comparison group is healthier at the onset of the study. It should come as no surprise that treatments in such studies tend to appear harmful. However, if health prior to treatment is adequately captured by H, for example if $H = 0$ denotes unhealthy and $H = 1$ denotes healthy, then we might assume that, conditional on H, whether or not $A = 1$ is random. We might have that $P(A = 1|H = 0) = 0.7$ and $P(A = 1|H = 1) = 0.4$, but we can adjust for this in our analysis using the statistical solution to the FPCI. The conditional independence at (3.2) implies the mean independence $E(Y(0)|H) = E(Y(0)|H, A = 0)$, and the consistency assumption implies that $E(Y(0)|H, A = 0) = E(Y|H, A = 0)$, and thus we can estimate $E(Y(0)|H)$ with $\hat{E}(Y|H, A = 0)$. Note that the positivity assumption guarantees that in a large enough sample, there will be individuals in the $(H, A = 0)$ stratum to inform this estimate. Similarly, we can estimate $E(Y(1)|H)$ with $\hat{E}(Y|H, A = 1)$. Therefore we can contrast the conditional averages $E(Y(1)|H)$ with $E(Y(0)|H)$.

We illustrate with an analysis of the What-If? study to attempt to address whether reducing drinking $A = 1$ reduces unsuppressed viral load $Y = 1$ at 4 months. Note that we could randomize participants to receive treatment with naltrexone $T = 1$, but randomizing them to actually reduce drinking $A = 1$ is not as feasible. We will let H indicate unsuppressed viral load at baseline. We assume

$$(Y(0), Y(1)) \amalg A|H.$$

From Table 1.3 of chapter 1, we have

```
> xtabs(~Y+A+H,data=whatifdat)
, , H = 0

   A
Y    0  1
  0 30 63
  1  6  7

, , H = 1

   A
Y    0  1
  0  6 13
  1 18 22
```

We thus estimate $E(Y(0)|H=0)$ by $\hat{E}(Y|H=0,A=0)=6/36=0.167$, and $E(Y(1)|H=0)$ by $\hat{E}(Y|H=0,A=1)=7/70=0.100$. Thus for the group with suppressed viral load at baseline, reducing drinking leads to a reduction in the chance of unsuppressed viral load at 4 months from 16.7% to 10%. Similarly we estimate $E(Y(0)|H=1)$ by $\hat{E}(Y|H=1,A=0)=18/24=0.750$, and $E(Y(1)|H=1)$ by $\hat{E}(Y|H=1,A=1)=22/35=0.629$. Thus for the group with unsuppressed viral load at baseline, reducing drinking leads to a reduction in the chance of unsuppressed viral load at 4 months from 75% to 62.9%.

We have shown how to use the conditional independence assumption at (3.2) to estimate conditional causal effects, such as $E(Y(1)|H)$ versus $E(Y(0)|H)$. In Chapter 6, we will explain how to use it to estimate unconditional causal effects, such as $E(Y(1))$ versus $E(Y(0))$, and also a special kind of conditional causal effect, called the *average effect of treatment on the treated* (ATT), i.e. $E(Y(1)|A=1)$ versus $E(Y(0)|A=1)$. Note that if A were a completely randomized treatment as in (3.1), then the ATT would equal the overall treatment effect, $E(Y(1))$ versus $E(Y(0))$, because A would be independent of the potential outcomes. We will also use the method of differences in differences introduced in Chapter 7 and the method of instrumental variables introduced in Chapter 9 to estimate the ATT. Finally, we note that one can also estimate the effect of treatment on the untreated using the same methods.

3.3 Effect Measures

Once we have estimated $E(Y(0))$ and $E(Y(1))$, there are several measures we can use to contrast them. For binary Y, $E(Y(0))$ and $E(Y(1))$ are probabilities, sometimes referred to as *risks*. For ease of exposition, let $p_1 = E(Y(1))$ and $p_0 = E(Y(0))$. The ratios $p_1/(1-p_1)$ and $p_0/(1-p_0)$ are called *odds*.

Given a risk p, we can compute the corresponding odds $o = p/(1-p)$, and given the odds o, we can compute the risk $p = o/(1+o)$. Arguably, it is more natural to think in terms of risks rather than odds; it is probably easier to understand that the risk of 28-day survival is 75.4% than it is to understand that the odds are 3.065. However, sometimes odds are natural, such as in betting, where 2:1 odds (or $o = 2$) represents a risk (i.e. chance of winning) of 0.667, and 1:1 odds ($o = 1$) represents a risk of 0.5. For binary Y, the most common effect measures are the *risk difference* (RD), *relative risk* (RR), and *odds ratio* (OR). Less common but also useful is what we will call the *other relative risk*, denoted by RR*, first brought to our attention by J.P. Scanlan (see Scanlan (2006)). These four measures are defined as

$$\begin{aligned} \text{RD} &= p_1 - p_0 \\ \text{RR} &= \frac{p_1}{p_0} \\ \text{RR*} &= \frac{1-p_0}{1-p_1} \\ \text{OR} &= \frac{p_1}{1-p_1} \Big/ \frac{p_0}{1-p_0} \end{aligned}$$

In the COVID-19 example, we estimated p_1 at 0.823 and p_0 at 0.754, therefore we have

$$\begin{aligned} \hat{\text{RD}} &= 0.069 \\ \hat{\text{RR}} &= 1.092 \\ \hat{\text{RR}}\text{*} &= 1.390 \\ \hat{\text{OR}} &= 1.293 \end{aligned}$$

These four measures always agree, qualitatively, in terms of the direction of the change in risk. In the COVID-19 example, dexamethasone increases the risk of 28-day survival from 0.754 to 0.823, and so the risk difference is positive, while the other three measures are greater than one. The other relative risk, RR*, deserves some explanation. Suppose the relative risk is greater than one, e.g. dexamethasone increases 28-day survival relative to withholding it. Then equivalently, the other relative risk must also be greater than one, e.g. withholding dexamethasone increases 28-day mortality relative to administering it.

We can also use the RD, RR, RR*, and OR to measure association between two binary variables such as Y and T. We simply let $p_1 = E(Y|T=1)$ and $p_0 = E(Y|T=0)$. When the randomized clinical trial assumption (3.1) holds, the measures of association and the measures of causal effect are equal. When we need to distinguish the causal measures from the association measures, we will refer to the causal measures as causal RD, causal RR, causal RR*, and causal OR, whereas RD, RR, RR*, and OR will refer to the association measures.

There are other causal effect measures for binary outcomes that are important to consider, however, they can each be written as a function of just one of the RD, RR, or RR*. The *number needed to treat* (NNT) (see Cook and Sackett (1995)) is

$$\frac{1}{p_1 - p_0} = \frac{1}{RD}.$$

This measures the number of individuals we need to treat in order to benefit one. Suppose we treat a sample of size N. Then Np_1 is the number who we would expect to survive 28 days. Had we not treated that same sample, then we would expect only Np_0 to survive 28 days. The number needed to treat is the value for N that makes the difference $Np_1 - Np_0$ equal to one, i.e. $NNT(p_1 - p_0) = 1$, or $NNT = 1/(p_1 - p_0)$. We estimate that we would need to treat $1/0.069 = 15$ COVID-19 patients with dexamethasone in order to benefit one. Note that even though $1/0.069 = 14.49$, we did not round down to 14, because then NNT would be slightly less than one. Number needed to treat is a measure useful in prioritizing scarce resources. For example, the NNT for a vaccine preventing severe COVID-19 is less for the 65+ group than it will be for essential workers, which is why some states prioritized the 65+ group first.

Three other causal effect measures that we will consider require us to assume that treatment cannot harm anyone; that is, there is no one of causal type 'harmed'. The *attributable fraction among the exposed* (Rothman et al. (2008)), which we will simply call the *attributable fraction* (AF), and which is also called the *probability of necessity* (PN) (Pearl (1999)), is

$$(p_1 - p_0)/p_1 = 1 - 1/RR.$$

Suppose we treat a sample of size N, then, once again, Np_1 is the number we would expect to survive 28 days. Further, $Np_1 - Np_0$ is the number of those for whom treatment was necessary for survival. Therefore, the fraction of those who were treated and survived whose survival we could attribute to treatment (i.e. for whom treatment was necessary), is

$$N(p_1 - p_0)/Np_1 = (p_1 - p_0)/p_1.$$

The probability of necessity is the conditional probability that treatment was necessary for suvival given treatment was administered and the individual survived. For the COVID-19 data, $PN = 0.084$. The attributable fraction is useful in legal contexts. For example, suppose it is claimed that a teacher died from COVID-19 due to teaching in person. The attributable fraction would measure the chance that the death was due to teaching in person, versus other activities.

The *causal power* (CP) (Cheng (1997)), also called the *probability of sufficiency* (PS) (Pearl (1999)), is

$$(p_1 - p_0)/(1 - p_0) = 1 - 1/\text{RR}^*.$$

Suppose we withhold treatment from a sample of size N; then, $N(1-p_0)$ is the number we would expect to die within 28 days. Further, $N(1-p_0)-N(1-p_1)$ is the number of those we would expect to have survived for 28 days had we treated them. Therefore, the fraction of those who were not treated and died within 28 days for whom treatment would have had the power to cause 28-day survival (i.e. for whom treatment would have been sufficient), is

$$\frac{N(1-p_0)-N(1-p_1)}{N(1-p_0)} = \frac{p_1 - p_0}{1-p_0}.$$

The probability of sufficiency is the conditional probability that treatment would have been sufficient for survival given treatment was withheld and the individual died. For the COVID-19 data, $PS = 0.280$. The causal power measures the power of a treatment to cause an outcome. For example, suppose a person did not receive a COVID-19 vaccine and subsequently got COVID-19. The causal power measures the chance that had that person been given the vaccine, COVID-19 would have been prevented.

Finally, the *probability of necessity and sufficiency* (PNS) (Pearl (1999)) is

$$p_1 - p_0 = RD.$$

We note that if we treat a sample of size N, then we expect $N(p_1-p_0)$ to need treatment for survival, and $N(1-p_0)-N(1-p_1) = N(p_1-p_0)$ is also the number for whom we would expect treatment is sufficient for preventing death. Therefore $N(p_1-p_0)/N$ is the unconditional probability that treatment is both necessary and sufficient for survival (e.g. it is the proportion of 'responsive' individuals). For the COVID-19 data, $PNS = 0.069$. We note that PNS will always be less than both PN and PS. Due to its dependence on the risk difference, PNS goes hand-in-hand with NNT.

The effect measures we have introduced to measure unconditional causal effects can also measure conditional causal effects; just repurpose p_1 and p_0 to denote conditional probabilities, such as $E(Y(1)|H=h)$ and $E(Y(0)|H=h)$. The parametric models introduced in Chapter 2 are useful for estimating both unconditional and conditional causal effects. Indeed, the linear model has a natural correspondence with the risk difference, whereas the loglinear and logistic models have natural correspondences with the relative risk and odds ratio, respectively. For unconditional causal effects, suppose the randomized clinical trial assumption (3.1) holds so that $E(Y|T=1) = E(Y(1))$ and $E(Y|T=0) = E(Y(0))$. The models are

$$\begin{aligned} \text{Linear} \quad & E(Y|T) = \beta_1 + \beta_2 T \\ \text{Loglinear} \quad & \log(E(Y|T)) = \beta_1 + \beta_2 T \\ \text{Logistic} \quad & \text{logit}(E(Y|T)) = \beta_1 + \beta_2 T \end{aligned} \tag{3.4}$$

For the linear model, $\beta_2 = E(Y|T=1) - E(Y|T=0) = RD$. For the loglinear model, $\beta_2 = \log(E(Y|T=1)) - \log(E(Y|T=0)) = \log(RR)$, so that $\exp(\beta_2) = RR$. For the logistic model, $\beta_2 = \text{logit}(E(Y|T = 1)) - \text{logit}(E(Y|T = 0))$, and $\text{logit}(p) = \log(o)$ where p is a risk and o is the odds $p/(1-p)$ corresponding to that risk. Therefore,

$$\beta_2 = \log \frac{E(Y|T = 1)}{1 - E(Y|T = 1)} - \log \frac{E(Y|T = 0)}{1 - E(Y|T = 0)} = \log(OR),$$

because

$$\log(o_1) - \log(o_0) = \log(o_1/o_0),$$

where o_1 and o_0 are two odds. In general, the $\log(\cdot)$ function has the properties that

$$\log(ab) = \log(a) + \log(b)$$

and

$$\log(a/b) = \log(a) - \log(b).$$

Thus for the logistic model, $\exp(\beta_2) = OR$.

We illustrate with the General Social Survey dataset, `gss`, introduced in Chapter 1. We will use these data to compare unconditional association measures with conditional effect measures assuming (3.2) holds but (3.1) does not. That (3.2) holds is a dubious assumption in this example, which we make for expository purposes only. The estimates of conditional effect measures can also be viewed as estimates of conditional association measures, which are sometimes called *adjusted* associations. The `gss` dataset has some variables with missing data. Because we are comparing unconditional measures with conditional measures, which require us to use more variables, we first restrict the dataset to participants with complete data on all of the variables we will use. This excludes 180 participants out of 2348, or 7.67%. Then the function `bootu.r`, with a 'u' for unconditional or unadjusted, computes estimates of the unconditional association measures and their 95% bootstrap confidence intervals. Note that the `glm` function with the `family=gaussian`, `family=poisson`, and `family=binomial` options solve the estimating equations of Chapter 2 for the parameters of the linear, loglinear, and logistic models, respectively. We certainly do not believe that our binary outcomes satisfy the distributional assumptions for the gaussian and the poisson models, but we are only using the `glm` function to solve estimating equations, and not to compute confidence intervals (we use the bootstrap for the latter).

We use the software to estimate the probability of voting for Trump conditional on an educational background of completing more than high school.

```
> gssr <-
  gss[, c("trump", "gthsedu", "magthsedu", "white", "female", "gt65")]
> gssrcc <- gssr[complete.cases(gssr), ]
bootu.r <- function ()
{
  # estimate the conditional probabilities
  # and the four association measures
```

```
estimator <- function(data, ids)
{
  dat <- data[ids,]
  # estimate the conditional probabilities
  mod <- glm(trump ~ gthsedu, family = gaussian, data = dat)
  p0 <- mod$coef[1]
  p1 <- mod$coef[1] + mod$coef[2]
  # estimate the association measures
  rd <- mod$coef[2]
  # use a loglinear model to estimate the log relative risk
  logrr <-
    glm(trump ~ gthsedu, family = poisson, data = dat)$coef[2]
  # prepare to estimate the log other relative risk
  trumpstar <- 1 - dat$trump
  gthsedustar <- 1 - dat$gthsedu
  # use a loglinear model to estimate the log other relative risk
  logrrstar <-
    glm(trumpstar ~ gthsedustar,
        family = poisson,
        data = dat)$coef[2]
  logor <-
    glm(trump ~ gthsedu, family = binomial, data = dat)$coef[2]
  # return the measures
  c(p0, p1, rd, logrr, logrrstar, logor)
}
boot.out <- boot(data = gssrcc,
                 statistic = estimator,
                 R = 1000)
# estimate the conditional probabilities and the
# confidence intervals
p0hat <- summary(boot.out)$original[1]
p0lci <- p0hat - 1.96 * summary(boot.out)$bootSE[1]
p0uci <- p0hat + 1.96 * summary(boot.out)$bootSE[1]
p1hat <- summary(boot.out)$original[2]
p1lci <- p1hat - 1.96 * summary(boot.out)$bootSE[2]
p1uci <- p1hat + 1.96 * summary(boot.out)$bootSE[2]
# estimate the association measures and the
# confidence intervals
rdhat <- summary(boot.out)$original[3]
logrrhat <- summary(boot.out)$original[4]
logrrstarhat <- summary(boot.out)$original[5]
logorhat <- summary(boot.out)$original[6]
rdhatlci <- rdhat - 1.96 * summary(boot.out)$bootSE[3]
rdhatuci <- rdhat + 1.96 * summary(boot.out)$bootSE[3]
logrrhatlci <- logrrhat - 1.96 * summary(boot.out)$bootSE[4]
logrrhatuci <- logrrhat + 1.96 * summary(boot.out)$bootSE[4]
logrrstarlci <- logrrstarhat - 1.96 * summary(boot.out)$bootSE[5]
logrrstaruci <- logrrstarhat + 1.96 * summary(boot.out)$bootSE[5]
logorhatlci <- logorhat - 1.96 * summary(boot.out)$bootSE[6]
```

```
  logorhatuci <- logorhat + 1.96 * summary(boot.out)$bootSE[6]
  rrhat <- exp(logrrhat)
  rrstarhat <- exp(logrrstarhat)
  orhat <- exp(logorhat)
  p1ci <- c(p1lci, p1uci)
  p0ci <- c(p0lci, p0uci)
  rdci <- c(rdhatlci, rdhatuci)
  rrci <- exp(c(logrrhatlci, logrrhatuci))
  rrstarci <- exp(c(logrrstarlci, logrrstaruci))
  orci <- exp(c(logorhatlci, logorhatuci))
  # return all the estimates
  list(
    p1hat = p1hat,
    p1ci = p1ci,
    p0hat = p0hat,
    p0ci = p0ci,
    rdhat = rdhat,
    rdci = rdci,
    rrhat = rrhat,
    rrci = rrci,
    rrstarhat = rrstarhat,
    rrstarci = rrstarci,
    orhat = orhat,
    orci = orci
  )
}
> bootu.out<-bootu.r()
> bootu.out
$p1hat
[1] 0.27117
$p1ci
[1] 0.24095 0.30139
$p0hat
[1] 0.23338
$p0ci
[1] 0.21030 0.25647
$rdhat
[1] 0.037782
$rdci
[1] 0.00078085 0.07478354
$rrhat
[1] 1.1619
$rrci
[1] 1.0047 1.3437
$rrstarhat
[1] 1.0518
$rrstarci
[1] 1.0006 1.1057
$orhat
```

```
[1] 1.2221
$orci
[1] 1.0056 1.4853
```

Letting Y indicate a reported vote for Trump and T indicate more than high school education, we find that the estimate and 95% confidence interval for $E(Y|T = 1)$ are $0.271(0.241, 0.301)$, whereas those for $E(Y|T = 0)$ are $0.233(0.210, 0.256)$. Therefore, individuals with more than high school education appear more likely to vote for Trump. As we are not assuming the independence at (3.1), we cannot state that more education causes a vote for Trump. Estimates of our four association measures with their 95% confidence intervals are presented in Table 3.2.

TABLE 3.2
Four Association Measures Relating More than High School Education to Voting for Trump

Measure	Estimate	95% Confidence Interval
RD	0.038	(0.001, 0.0748)
RR	1.16	(1.005, 1.34)
RR*	1.05	(1.001, 1.11)
OR	1.22	(1.006, 1.49)

We observe that the confidence intervals just barely exclude the null association, which is 0 for the RD and 1 for the other three measures. In fact, for a few executions of the program, the confidence intervals very slightly did not exclude the null associations. Thus, the estimated association between education and voting for Trump is on the boundary of statistical significance.

Next, we estimate the conditional effect measures under assumption (3.2) with H = (`magthsedu`, `white`, `female`, `gt65`) and A replaced by T. We did not incorporate father's education, `pagthsedu`, because it is missing for 583 respondents. The models are

$$\begin{aligned} \text{Linear} \quad & E(Y|T) = \beta_1 + \beta_2 T + \beta_3 H_1 + \cdots + \beta_6 H_4 \\ \text{Loglinear} \quad & \log(E(Y|T)) = \beta_1 + \beta_2 T + \beta_3 H_1 + \cdots + \beta_6 H_4 \\ \text{Logistic} \quad & \text{logit}(E(Y|T)) = \beta_1 + \beta_2 T + \beta_3 H_1 + \cdots + \beta_6 H_4 \end{aligned} \tag{3.5}$$

Because these are conditional measures, we need to specify values for $H_1, \ldots, H_4$. Some researchers set these equal to their averages, $E(H_1), \ldots, E(H_4)$, estimated from the data. However, for binary variables, this is awkward. We set them all equal to one. Letting $H = (H_1, H_2, H_3, H_4)$, we denote $H = (1, 1, 1, 1)$ simply as $H = 1$. Therefore, we are estimating the conditional effect for a white female participant who is older than 65 and whose mother has more than a high school education.

Again, for the linear model,

$$\beta_2 = E(Y|T = 1, H = 1) - E(Y|T = 0, H = 1) = RD.$$

For the loglinear model,

$$\beta_2 = \log(E(Y|T = 1, H = 1)) - \log(E(Y|T = 0, H = 1)) = \log(RR),$$

so that $\exp(\beta_2) = RR$. For the logistic model,

$$\beta_2 = \text{logit}(E(Y|T = 1, H = 1)) - \text{logit}(E(Y|T = 0, H = 1)),$$

so that

$$\beta_2 = \log \frac{E(Y|T = 1, H = 1)}{1 - E(Y|T = 1, H = 1)} - \log \frac{E(Y|T = 0, H = 1)}{1 - E(Y|T = 0, H = 1)} = \log(OR),$$

and $\exp(\beta_2) = OR$.

Because our outcome is binary, we choose to fit the logistic parametric model. We do not object to computing the sampling variability of our estimates assuming the Binomial distribution for our outcomes, and therefore we present the estimated standard errors and p-values returned by `glm`. We observe that all of the coefficient estimates are statistically significant.

```
> lmoda <-
  glm(trump ~ gthsedu + magthsedu + white + female + gt65,
      family = binomial,
      data = gssrcc)
> summary(lmoda)
Call:glm(
  formula = trump ~ gthsedu + magthsedu + white + female +
    gt65,
  family = binomial,
  data = gssrcc
)
Coefficients:
            Estimate Std. Error z value Pr(>|z|)
(Intercept)   -2.913      0.212  -13.73  < 2e-16
gthsedu        0.232      0.111    2.08    0.038
magthsedu     -0.573      0.137   -4.18  2.9e-05
white          2.352      0.209   11.25  < 2e-16
female        -0.513      0.106   -4.82  1.4e-06
gt65           0.572      0.121    4.73  2.2e-06
```

Next we use the bootstrap to estimate the sampling distributions for $p_1 = E(Y|T = 1, H = 1)$, $p_0 = E(Y|T = 0, H = 1)$, and our four effect measures.

```
lmodboot.r <- function ()
{
  # estimate the expected conditional potential outcomes
  # and the conditional effect or association measures
```

```
estimator <- function(data, ids)
{
  dat <- data[ids, ]
  coef <-
    glm(
      trump ~ gthsedu + magthsedu + white + female + gt65,
      family = binomial,
      data = dat
    )$coef
  xbeta1 <- sum(coef)
  xbeta0 <- sum(coef) - coef[2]
  p1 <- exp(xbeta1) / (1 + exp(xbeta1))
  p0 <- exp(xbeta0) / (1 + exp(xbeta0))
  rd <- p1 - p0
  logrr <- log(p1) - log(p0)
  logrrstar <- log((1 - p0)) - log((1 - p1))
  logor <- log(p1 / (1 - p1)) - log(p0 / (1 - p0))
  c(p1, p0, rd, logrr, logor, logrrstar)
}
boot.out <- boot(data = gssrcc,
                 statistic = estimator,
                 R = 1000)
# extract the estimates and the confidence intervals as before
p1hat <- summary(boot.out)$original[1]
p0hat <- summary(boot.out)$original[2]
rdhat <- summary(boot.out)$original[3]
logrrhat <- summary(boot.out)$original[4]
logorhat <- summary(boot.out)$original[5]
logrrstarhat <- summary(boot.out)$original[6]
p1hatlci <- p1hat - 1.96 * summary(boot.out)$bootSE[1]
p1hatuci <- p1hat + 1.96 * summary(boot.out)$bootSE[1]
p0hatlci <- p0hat - 1.96 * summary(boot.out)$bootSE[2]
p0hatuci <- p0hat + 1.96 * summary(boot.out)$bootSE[2]
rdhatlci <- rdhat - 1.96 * summary(boot.out)$bootSE[3]
rdhatuci <- rdhat + 1.96 * summary(boot.out)$bootSE[3]
logrrhatlci <- logrrhat - 1.96 * summary(boot.out)$bootSE[4]
logrrhatuci <- logrrhat + 1.96 * summary(boot.out)$bootSE[4]
logorhatlci <- logorhat - 1.96 * summary(boot.out)$bootSE[5]
logorhatuci <- logorhat + 1.96 * summary(boot.out)$bootSE[5]
logrrstarlci <- logrrstarhat - 1.96 * summary(boot.out)$bootSE[6]
logrrstaruci <- logrrstarhat + 1.96 * summary(boot.out)$bootSE[6]
rrhat <- exp(logrrhat)
orhat <- exp(logorhat)
rrstarhat <- exp(logrrstarhat)
p1ci <- c(p1hatlci, p1hatuci)
p0ci <- c(p0hatlci, p0hatuci)
rdci <- c(rdhatlci, rdhatuci)
rrci <- exp(c(logrrhatlci, logrrhatuci))
orci <- exp(c(logorhatlci, logorhatuci))
```

```
  rrstarci <- exp(c(logrrstarlci, logrrstaruci))
  # return the estimates
  list(
    p1hat = p1hat,
    p1ci = p1ci,
    p0hat = p0hat,
    p0ci = p0ci,
    rdhat = rdhat,
    rdci = rdci,
    rrhat = rrhat,
    rrci = rrci,
    rrstarhat = rrstarhat,
    rrstarci = rrstarci,
    orhat = orhat,
    orci = orci
  )
}
> lmodboot.out<-lmodboot.r()
> lmodboot.out
$p1hat
[1] 0.30101
$p1ci
[1] 0.22995 0.37208
$p0hat
[1] 0.25458
$p0ci
[1] 0.18518 0.32397
$rdhat
[1] 0.046438
$rdci
[1] 0.0022706 0.0906057
$rrhat
[1] 1.1824
$rrci
[1] 1.0039 1.3927
$rrstarhat
[1] 1.0664
$rrstarci
[1] 1.0018 1.1352
$orhat
[1] 1.261
$orci
[1] 1.0091 1.5758
```

We find that the estimate and 95% confidence interval for $E(Y|T = 1, H = 1)$ are $0.301(0.230, 0.372)$, whereas those for $E(Y|T = 0, H = 1)$ are $0.255(0.185, 0.324)$. Therefore, white female participants who are older than 65 years and whose mothers have more than a high school education appear more likely to vote for Trump if they themselves have more than a high schol education. If we were to assume the independence at (3.2), we could state that more education causes a vote for Trump for women in this category. Estimates of our four association or effect measures with their 95% confidence intervals are presented in Table 3.3.

TABLE 3.3
Four Conditional Association or Effect Measures Relating More than High School Education to Voting for Trump

Measure	Estimate	95% Confidence Interval
RD	0.046	(0.002, 0.091)
RR	1.18	(1.004, 1.39)
RR*	1.07	(1.002, 1.14)
OR	1.26	(1.009, 1.58)

We observe that the confidence intervals exclude the null association, which is 0 for the RD and 1 for the other three measures, although just slightly. Thus, the estimated conditional association between more than high school education and voting for Trump is on the boundary of statistical significance.

For non-binary outcomes, we can use one of the three parametric models at (3.4) or (3.5) with adjusted interpretations of effect, conditional effect, association, or conditional association. For example,

$$\begin{array}{rc} \text{Linear} & \beta_2 = E(Y(1)) - E(Y(0)) \\ \text{Loglinear} & \exp(\beta_2) = \frac{E(Y(1))}{E(Y(0))} \\ \text{Logistic} & \beta_2 = \log \frac{E(Y(1))}{1-E(Y(1))} - \log \frac{E(Y(0))}{1-E(Y(0))} \end{array}$$

For the linear model, $\beta_2 = b > 0$ corresponds to treatment increasing the average potential outcome by b units. For the loglinear model, $\exp(\beta_2) = b > 0$ corresponds to treatment increasing the average potential outcome by a factor of b. For the logistic model, $\beta_2 = b > 0$ corresponds to treatment increasing the logit of the average potential outcome by b. If β_2 is negative, suitable adjustments to the interpretation can be made. The choice of parametric model should respect the range of the data. For example, for the logistic model, outcomes outside the range of [0,1] must either be shrunk to fit that range, or another one of the models should be chosen.

3.4 Exercises

1. Construct a complete case dataset from `gss` with the variables `owngun`, `conservative`, `gt65`, `female`, and `white`. Use these data to estimate the unconditional probabilities and four unconditional association measures analogous to those in Table 3.2 relating conservative political views to owning a gun. Next, estimate the conditional probabilities and four conditional association or effect measures analogous to those in Table 3.3 relating conservative political views to owning a gun, conditional on being a male who is greater than 65 years and self-reported as white. For the purposes of this exercise, assume that conservative political views are formed before any gun purchases. Is this assumption realistic? What assumptions would be needed to interpret these estimates causally?

2. Using the `gss` dataset, create a new binary variable called `pamiss` that is one if `pagthsedu` is missing (`NA`), and then recode the missing values of `pagthsedu` to equal zero. Then construct a complete case dataset from `gss` with the variables `trump`, `gthsedu`, `magthsedu`, the recoded `pagthsedu`, `pamiss`, `white`, `female`, and `gt65`. Use this dataset to re-estimate the conditional probabilities and associations or effect measures of Table 3.3 relating more than high school education to voting for Trump, now additionally adjusting for father's education by including `pagthsedu` and `pamiss` in the parametric logistic model as covariates and setting `pamiss` to 0 and `pagthsedu` to 1. This is one approach to incorporating a covariate that is missing for some participants. How do these estimates differ from those of Table 3.3?

3. The dataset `brfss` is constructed from the 2019 Behavorial Risk Factor Surveillance System (BRFSS) survey data, available at Centers for Disease Control and Prevention (2020). The BRFSS is a health-related telephone survey that collects data about U.S. residents aged 18 and over regarding their health-related risk behaviors, chronic health conditions, and use of preventive services. The binary variables included in the complete case dataset `brfss` are `gt65` recording age greater than 65, `female` recording female, `whitenh` recording self-identification as white non-hispanic, `blacknh` recording self-identification as black non-hispanic, `hisp` recording self-identification as hispanic, `othernh` recording self-identification as an other racial category and non-hispanic, `multinh` recording self-identification as multi-racial and non-hispanic, `rural` recording residence in a rural county, `insured` recording having health insurance, `flushot` recording a flu shot within the past 12 months, `gthsedu` recording education greater than high school, and `zerodrinks`

recording zero alcholic drinks during the past 30 days. We also include the count variable `maxdrinks` recording the largest number of alcoholic drinks had on any occasion during the past 30 days.

```
> head(brfss)
  gt65 female whitenh blacknh hisp othernh multinh rural insured flushot gthsedu zerodrinks maxdrinks
1    1      1       0       1    0       0       0     0       1       0       0          1         0
2    1      1       1       0    0       0       0     0       1       1       1          1         0
3    1      1       0       1    0       0       0     0       1       1       1          1         0
4    1      1       1       0    0       0       0     1       1       0       1          1         0
5    1      0       1       0    0       0       0     1       1       1       1          0         3
6    1      0       1       0    0       0       0     0       1       1       1          0         1
```

We will use a two-part model to investigate the "effect" of residing in a rural county on alcohol consumption, conditional on confounders. First we model the probability of consuming zero alcoholic drinks, and second we model the expected largest number of alcoholic drinks on any occasion for those who consumed alcohol.

Using a logistic parametric model, estimate the conditional probabilities and association or effect measures relating residence in a rural county to consuming zero alcoholic drinks during the past 30 days, conditional on being male, white, non-hispanic, less than 65 years of age, and having had greater than high school education. Be sure to put all the binary variables regarding race and ethnicity into the parametric model except the one you choose as the reference category. What assumptions would be necessary to interpret this analysis causally?

Restrict the `brfss` dataset to only include survey respondents who consumed alcoholic drinks during the past 30 days. Using a log-linear parametric model, estimate the conditional expectations of the potential outcomes and the association or effect measures relating residence in a rural county to the largest number of alcoholic drinks had on any occasion during the past 30 days, conditional on being male, white, non-hispanic, less than 65 years of age, and having had greater than high school education. Be sure to put all the binary variables regarding race and ethnicity into the parametric model except the one you choose as the reference category. What assumptions would be necessary to interpret this analysis causally?

4. According to a Stat News report on November 18, 2020 (Stat News (2020)) the Covid-19 vaccine developed by Pfizer and BioNTech is estimated to have an efficacy of 95%. The study included 43,661 volunteers randomized to a vaccine versus placebo. In the placebo group, 162 developed Covid-19, whereas in the vaccine group, 8 developed Covid-19. Assuming approximately equal numbers of participants in each group were exposed to the coronavirus, use these data to validate the claim of 95% estimated efficacy. Hint: vaccine efficacy is a causal effect measure analogous to the attributable fraction. Suppose that the ratio of the number exposed to the coronavirus in the vaccine group to that in the placebo group were actually 0.8, due to chance. How would this change your estimate of the efficacy?

4

Effect-Measure Modification and Causal Interaction

In this section, we revisit the RECOVERY trial of dexamethasone for COVID-19 (RECOVERY Collaborative Group (2020)) and incorporate data on the *effect modifier* $M = 1$, indicating invasive mechanical ventilation prior to treatment; the data are presented in Table 4.1. The trial pre-specified five *subgroup analyses*, including comparing the mortality outcome Y across $T = 1$ and $T = 0$ within the $M = 1$ stratum. Table 4.2 presents estimates and 95% confidence intervals for the average potential outcomes $Y(0)$ and $Y(1)$ together with the four main effect measures RD, RR, RR*, and OR, for the $M = 0$ and $M = 1$ strata. The R code below shows how to compute the estimates and confidence intervals using the bootstrap. The variable M is called an effect-modifier because the effect measures might differ, or be modified, from $M = 0$ to $M = 1$. It is natural to wonder if the effect of the treatment is *stronger* in one stratum than in another. This is sometimes a difficult question to answer. We need to first select an effect measure, and then compare it across strata. Whereas the effect measures will always agree qualitatively within a single stratum (e.g. RD>0, RR>1, RR*>1, OR>1), one measure may present a treatment effect as stronger in a certain stratum, while another measure may present it as weaker in that same stratum.

TABLE 4.1
RECOVERY Trial: Subgroup Analysis

M	T	Y	n
0	0	0	787
0	0	1	2851
0	1	0	368
0	1	1	1412
1	0	0	278
1	0	1	405
1	1	0	86
1	1	1	238

DOI: 10.1201/9781003146674-4

4.1 Effect-Measure Modification and Statistical Interaction

Another possibility is that one effect measure will show no statistically significant difference across strata, whereas another measure will show a difference. The next section explores these phenomena in more depth, but the examples of this section partially illustrate them. Table 4.2, computed using `bootinside.r` and `boot.r`, measures modification either by a difference (for $E(Y(1))$, $E(Y(0))$, and $\hat{RD}$) or a ratio (for $\hat{RR}$, $\hat{RR}^*$, and $\hat{OR}$). We see that all four measures indicate that the effect of dexamethasone is stronger in the $M = 1$ stratum; the difference of risk differences is 0.132 (0.064, 0.200), the ratio of relative risks is 1.224 (1.108, 1.351), the ratio of other relative risks is 1.466 (1.150, 1.868), and the ratio of odds ratios is 1.794 (1.279, 2.514). Thus there is effect-measure modification by receipt of invasive mechanical ventilation, and it is consistent across all four measures. We use the terminology *effect-measure modification* instead of *effect modification* because presence or absence of effect modification (as well as its strength) can depend upon the chosen measure (Rothman et al. (2008)).

```
bootinside.r <- function (data, ids)
{
  dat <- data[ids, ]
  # subset only the untreated
  dat0 <- dat[dat$T == 0, ]
  # subset only the treated
  dat1 <- dat[dat$T == 1, ]
  # estimate the expected potential outcomes
  EY0.0 <- mean(dat0$Y[dat0$M == 0])
  EY0.1 <- mean(dat0$Y[dat0$M == 1])
  EY1.0 <- mean(dat1$Y[dat1$M == 0])
  EY1.1 <- mean(dat1$Y[dat1$M == 1])
  # estimate the effect measures
  RD.0 <- EY1.0 - EY0.0
  RD.1 <- EY1.1 - EY0.1
  logRR.0 <- log(EY1.0 / EY0.0)
  logRR.1 <- log(EY1.1 / EY0.1)
  logRRstar.0 <- log((1 - EY0.0) / (1 - EY1.0))
  logRRstar.1 <- log((1 - EY0.1) / (1 - EY1.1))
  logOR.0 <- logRR.0 + logRRstar.0
  logOR.1 <- logRR.1 + logRRstar.1
  EY0diff <- EY0.1 - EY0.0
  EY1diff <- EY1.1 - EY1.0
  RDdiff <- RD.1 - RD.0
  logRRdiff <- logRR.1 - logRR.0
  logRRstardiff <- logRRstar.1 - logRRstar.0
  logORdiff <- logOR.1 - logOR.0
  # return all of the estimates
```

```
  c(
    EY0.0,
    EY0.1,
    EY0diff,
    EY1.0,
    EY1.1,
    EY1diff,
    RD.0,
    RD.1,
    RDdiff,
    logRR.0,
    logRR.1,
    logRRdiff,
    logRRstar.0,
    logRRstar.1,
    logRRstardiff,
    logOR.0,
    logOR.1,
    logORdiff
  )
}
boot.r <- function ()
{
  # estimate bootstrap confidence intervals and return the point
          estimates EM.out <- boot(data = dat,
                 statistic = bootinside.r,
                 R = 1000)
  EM.est <- summary(EM.out)$original
  EM.expest <- exp(EM.est)
  EM.SE <- summary(EM.out)$bootSE
  EM.lci <- EM.est - 1.96 * EM.SE
  EM.explci <- exp(EM.lci)
  EM.uci <- EM.est + 1.96 * EM.SE
  EM.expuci <- exp(EM.uci)
  EM.est <- data.frame(EM.est)
  EM.expest <- data.frame(EM.expest)
  EM.lci <- data.frame(EM.lci)
  EM.explci <- data.frame(EM.explci)
  EM.uci <- data.frame(EM.uci)
  EM.expuci <- data.frame(EM.expuci)
  dimnames(EM.est)[[1]] <-
    c(
      "EY0.0",
      "EY0.1",
      "EY0diff",
      "EY1.0",
      "EY1.1",
      "EY1diff",
      "RD.0",
```

```
      "RD.1",
      "RDdiff",
      "logRR.0",
      "logRR.1",
      "logRRdiff",
      "logRRstar.0",
      "logRRstar.1",
      "logRRstardiff",
      "logOR.0",
      "logOR.1",
      "logORdiff"
    )
  dimnames(EM.expest)[[1]] <-
    c(
      "EY0.0",
      "EY0.1",
      "EY0diff",
      "EY1.0",
      "EY1.1",
      "EY1diff",
      "RD.0",
      "RD.1",
      "RDdiff",
      "logRR.0",
      "logRR.1",
      "logRRdiff",
      "logRRstar.0",
      "logRRstar.1",
      "logRRstardiff",
      "logOR.0",
      "logOR.1",
      "logORdiff"
    )
  dimnames(EM.lci)[[1]] <-
    c(
      "EY0.0",
      "EY0.1",
      "EY0diff",
      "EY1.0",
      "EY1.1",
      "EY1diff",
      "RD.0",
      "RD.1",
      "RDdiff",
      "logRR.0",
      "logRR.1",
      "logRRdiff",
      "logRRstar.0",
      "logRRstar.1",
```

```
    "logRRstardiff",
    "logOR.0",
    "logOR.1",
    "logORdiff"
  )
dimnames(EM.explci)[[1]] <-
  c(
    "EY0.0",
    "EY0.1",
    "EY0diff",
    "EY1.0",
    "EY1.1",
    "EY1diff",
    "RD.0",
    "RD.1",
    "RDdiff",
    "logRR.0",
    "logRR.1",
    "logRRdiff",
    "logRRstar.0",
    "logRRstar.1",
    "logRRstardiff",
    "logOR.0",
    "logOR.1",
    "logORdiff"
  )
dimnames(EM.uci)[[1]] <-
  c(
    "EY0.0",
    "EY0.1",
    "EY0diff",
    "EY1.0",
    "EY1.1",
    "EY1diff",
    "RD.0",
    "RD.1",
    "RDdiff",
    "logRR.0",
    "logRR.1",
    "logRRdiff",
    "logRRstar.0",
    "logRRstar.1",
    "logRRstardiff",
    "logOR.0",
    "logOR.1",
    "logORdiff"
  )
dimnames(EM.expuci)[[1]] <-
  c(
```

```
      "EY0.0",
      "EY0.1",
      "EY0diff",
      "EY1.0",
      "EY1.1",
      "EY1diff",
      "RD.0",
      "RD.1",
      "RDdiff",
      "logRR.0",
      "logRR.1",
      "logRRdiff",
      "logRRstar.0",
      "logRRstar.1",
      "logRRstardiff",
      "logOR.0",
      "logOR.1",
      "logORdiff"
    )
  list(
    EM.est = EM.est,
    EM.expest = EM.expest,
    EM.lci = EM.lci,
    EM.explci = EM.explci,
    EM.uci = EM.uci,
    EM.expuci = EM.expuci
  )
}
```

Effect-measure modification is closely tied to statistical interaction. For the linear model

$$E(Y|T, M) = \beta_0 + \beta_1 T + \beta_2 M + \beta_3 T * M,$$

we can compute that the interaction term β_3 equals the difference of risk differences

$$E(Y|T = 1, M = 1) - E(Y|T = 0, M = 1) -$$
$$(E(Y|T = 1, M = 0) - E(Y|T = 0, M = 0)).$$

For the loglinear model

$$\log(E(Y|T, M)) = \beta_0 + \beta_1 T + \beta_2 M + \beta_3 T * M,$$

β_3 is the log of the ratio of relative risks,

$$\log(E(Y|T = 1, M = 1)/E(Y|T = 0, M = 1)) -$$
$$\log(E(Y|T = 1, M = 0)/E(Y|T = 0, M = 0)).$$

TABLE 4.2
Effect-measure Modification in the RECOVERY Trial

Measure	$M = 0$	$M = 1$	Modification
$\hat{E}(Y(0)\|M)$ (95% CI)	0.784 (0.770, 0.797)	0.593 (0.556, 0.630)	−0.191 (−0.230, −0.152)
$\hat{E}(Y(1)\|M)$ (95% CI)	0.793 (0.773, 0.813)	0.735 (0.685, 0.784)	−0.059 (−0.113, −0.005)
$\hat{RD}$ (95% CI)	0.010 (−0.015, 0.034)	0.142 (0.078, 0.205)	0.132 (0.064, 0.200)
$\hat{RR}$ (95% CI)	1.012 (0.982, 1.044)	1.239 (1.127, 1.361)	1.224 (1.108, 1.351)
$\hat{RR}*$ (95% CI)	1.046 (0.931, 1.176)	1.533 (1.240, 1.896)	1.466 (1.150, 1.868)
$\hat{OR}$ (95% CI)	1.059 (0.914, 1.227)	1.900 (1.402, 2.573)	1.794 (1.279, 2.514)

Letting $\bar{T} = 1 - T$ and $\bar{Y} = 1 - Y$, the loglinear model

$$\log(E(\bar{Y}|\bar{T}, M)) = \beta_0 + \beta_1\bar{T} + \beta_2 M + \beta_3\bar{T} * M,$$

has β_3 as the log of the ratio of the other relative risks. Finally, for the logistic model,

$$\text{logit}(E(Y|T, M)) = \beta_0 + \beta_1 T + \beta_2 M + \beta_3 T * M,$$

β_3 is the log of the ratio of odds ratios.

Storing the dexamethasone data in the data frame `dat` with header

```
> head(dat)
  id M T Y Tbar Ybar
1  1 0 0 0    1    1
2  2 0 0 0    1    1
3  3 0 0 0    1    1
4  4 0 0 0    1    1
5  5 0 0 0    1    1
6  6 0 0 0    1    1
```

we can compute β_3 for each of these models using the `glm` function or the `gee` function in R. We have previously used the `glm` function, but the `gee` function, which is an acronym for the *generalize estimating equations* (Zeger et al. (1988)) that we use to compute estimates and large sample sampling distributions (with the *robust standard error* option), can give us a quick answer about whether the effect-measure modification is statistically significant. We look at the coefficient estimate divided by its robust standard error, which gives us the *robust z-statistic*. This tells us how many standard errors the coefficient estimate is away from zero. For example, for the linear model,

```
> linmod <- gee(Y ~ T + M + T * M,
                id = id,
                data = dat,
                family = gaussian)
> linmod
 GEE:  GENERALIZED LINEAR MODELS FOR DEPENDENT DATA
 gee S-function, version 4.13 modified 98/01/27 (1998)
Model:
 Link:                      Identity
 Variance to Mean Relation: Gaussian
 Correlation Structure:     Independent
Call:
gee(formula = Y ~ T + M + T * M, id = id, data = dat, family = gaussian)
Coefficients:
              Estimate Naive S.E.    Naive z Robust S.E.  Robust z
(Intercept)  0.7836723  0.0069757 112.34313   0.0068264 114.80019
T            0.0095861  0.0121702   0.78767   0.0117786   0.81386
M           -0.1907002  0.0175457 -10.86879   0.0199994  -9.53529
T:M          0.1320096  0.0308817   4.27468   0.0330741   3.99133
```

we see that β_3 is estimated as the coefficient of `T:M` (that is the way `R` represents interaction), or 0.132, and that this is 3.99 standard errors away from 0, which gives us a two-sided p-value of

```
> 2 * (1 - pnorm(3.99))
[1] 6.6073e-05
```

which is statistically signficant. For the loglinear model

```
> logmod
 GEE:  GENERALIZED LINEAR MODELS FOR DEPENDENT DATA
 gee S-function, version 4.13 modified 98/01/27 (1998)
Model:
 Link:                      Logarithm
 Variance to Mean Relation: Poisson
 Correlation Structure:     Independent
Call:
gee(formula = Y ~ T + M + T * M, id = id, data = dat, family = poisson)
Coefficients:
              Estimate Naive S.E.   Naive z Robust S.E.  Robust z
(Intercept) -0.243764  0.0091092 -26.76031   0.0087108 -27.98418
T            0.012158  0.0158278   0.76815   0.0149096   0.81545
M           -0.278844  0.0258282 -10.79610   0.0328768  -8.48146
T:M          0.201977  0.0427623   4.72324   0.0484000   4.17307
```

we have that β_3 is estimated at 0.202, which is 4.17 standard errors away from 0, hence also statistically significant. To find the ratio of relative risks, we calculate $\exp(\hat{\beta}_3) = 1.224$, again agreeing with Table 4.2. For the loglinear model

```
> logstarmod
 GEE:  GENERALIZED LINEAR MODELS FOR DEPENDENT DATA
 gee S-function, version 4.13 modified 98/01/27 (1998)
Model:
 Link:                      Logarithm
 Variance to Mean Relation: Poisson
 Correlation Structure:     Independent
Call:
gee(formula = Ybar ~ Tbar + M + Tbar * M, id = id, data = dat,
    family = poisson)
Coefficients:
              Estimate Naive S.E.   Naive z Robust S.E.  Robust z
(Intercept) -1.576286   0.045566 -34.59365    0.046428 -33.95092
Tbar         0.045325   0.055200   0.82109    0.056137   0.80739
M            0.249889   0.104693   2.38688    0.103427   2.41610
Tbar:M       0.382198   0.121160   3.15448    0.117583   3.25044
```

we have that $\hat{\beta}_3 = 0.382$, which is 3.25 standard errors away from 0, and is statistically signficant. To find the ratio of other relative risks, we calculate $\exp(\hat{\beta}_3) = 1.466$. Finally, for the logistic model

```
> logitmod
 GEE:  GENERALIZED LINEAR MODELS FOR DEPENDENT DATA
 gee S-function, version 4.13 modified 98/01/27 (1998)
Model:
 Link:                      Logit
 Variance to Mean Relation: Binomial
 Correlation Structure:     Independent
Call:
gee(formula = Y ~ T + M + T * M, id = id, data = dat, family = binomial)
Coefficients:
             Estimate Naive S.E.   Naive z Robust S.E.   Robust z
(Intercept)  1.287197   0.040279  31.95688    0.040267   31.96683
T            0.057483   0.071064   0.80888    0.071042    0.80913
M           -0.910931   0.087707 -10.38611    0.087679  -10.38935
T:M          0.584175   0.164194   3.55783    0.164143    3.55894
```

we have that $\exp(\hat{\beta}) = \exp(0.584) = 1.794$.

We conclude that dexamethasone has a stronger effect for patients who require invasive mechanical ventilation. In fact, dexamethasone is not statistically signficantly better than placebo for patients who do not require invasive mechanical ventilation, judging from the estimates and standard errors for the coefficient of T in the four models.

Next we turn to the NCES data from Chapter 1. We let M be `selective`, T be `female`, and Y be `highmathsat`. Using R code that is virtually identical to that for the dexamethasone example, we compute the estimates and 95% confidence intervals shown in Table 4.3. Using the risk difference, we see that the effect of admitting proportionally more women is to decrease the percentage of schools with high math SAT scores for both selective and non-selective schools. By necessity, the other three measures show the same thing, qualitatively. However, the risk difference suggests that the effect of admitting proportionally more women is stronger in the selective schools (difference of risk differences equal to $-0.244(-0.379, -0.109)$, while only the other relative risk agrees qualitatively with this assessment (ratio of other relative risks equal to $0.547(0.409, 0.731)$. Both the relative risk (ratio equal to $1.052(0.640, 1.729)$) and the odds ratio (ratio equal to $0.576(0.281, 1.180)$) suggest that the effect of admitting proportionally more women is not statistically significantly stronger or weaker for selective versus non-selective schools.

Note that a causal interpretation of the NCES results requires the stratified randomized trial assumption $(Y(0), Y(1)) \amalg T|M$; that is, potential math SAT scores to admitting proportionally more women are independent of actually admitting proportionally more women, conditional on a school's selectivity. This would be violated, for example, if among selective schools, those who tend to admit proportionally more women have an applicant pool with lower math SAT scores. While it is unlikely that conditioning on selectivity is enough to control for confounding, it likely that selectivity is an important confounder as well as an important effect-modifier. For the dexamethasone data, we automatically have $(Y(0), Y(1)) \amalg T|M$ due to randomization of T, which makes a causal interpretation entirely appropriate.

TABLE 4.3
Effect-measure Modification with the NCES Data

Measure	$M = 0$	$M = 1$	Modification
$\hat{E}(Y(0)\|M)$ (95% CI)	0.167 (0.135, 0.198)	0.675 (0.602, 0.748)	0.509 (0.430, 0.587)
$\hat{E}(Y(1)\|M)$ (95% CI)	0.081 (0.056, 0.106)	0.345 (0.242, 0.449)	0.264 (0.157, 0.371)
$\hat{RD}$ (95% CI)	−0.086 (−0.125, −0.046)	−0.330 (−0.460, −0.200)	−0.244 (−0.379, −0.109)
$\hat{RR}$ (95% CI)	0.486 (0.337, 0.700)	0.511 (0.366, 0.713)	1.052 (0.640, 1.729)
$\hat{RR}*$ (95% CI)	0.907 (0.866, 0.950)	0.496 (0.372, 0.661)	0.547 (0.409, 0.731)
$\hat{OR}$ (95% CI)	0.440 (0.293, 0.662)	0.254 (0.140, 0.458)	0.576 (0.281, 1.180)

4.2 Qualitative Agreement of Effect Measures in Modification

Brumback and Berg (2008) and Shannin and Brumback (2021) investigated conditions under which subsets of the four measures agree, qualitatively. Previously, Rothman and Greenland (1998) had shown that if both the effect modifier and the treatment have non-zero effects on the outcome, then (a) when there is no modification of the risk difference, there must be modification of the relative risk, (b) when there is no modification of the relative risk, there must be modification of the risk difference, and (c) when there is no modificiation of the odds ratio, there must be modification of the relative risk. Perhaps the most useful result concerning qualitative agreement is from Shannin and Brumback (2021), who prove that when the relative risk and other relative risk both change in the same direction (e.g. both increase) from one stratum of the modifier to another, then so must the risk difference and odds ratio. The proof is not difficult. Letting $E(Y|T=1,M=0)=p_1$, $E(Y|T=0,M=0)=p_0$, $E(Y|T=1,M=1)=p_3$, and $E(Y|T=0,M=1)=p_2$, we wish to show that $RR_0=p_1/p_0<RR_1=p_3/p_2$ and $RR_0^*=(1-p_0)/(1-p_1)<RR_1^*=(1-p_2)/(1-p_3)$ imply $RD_0=p_1-p_0<RD_1=p_3-p2$ and $OR_0=(p_1(1-p0))/(p_0(1-p_1))<OR_1=(p_3(1-p_2))/(p_2(1-p_3))$. That $OR_0<OR_1$ follows from $OR_0=RR_0RR_0^*$ and $OR_1=RR_1RR_1^*$, so that when the factors RR_0 and RR_0^* both increase, so necessarily does the product OR_0. To show that $RD_0<RD_1$, it suffices to show that $RR_0^*<RR_1^*$ implies that $p_3-p_2>(p_1-p_0)(1-p2)/(1-p_0)$ and that $RR_0<RR_1$ implies that $p_3-p_2>(p_1-p_0)(p_2/p_0)$; then, because $(1-p_2)/(1-p_0)$ and p_2/p_0 are on opposite sides of one, p_3-p_2 must be greater than p_1-p_0.

Arguably, the three most important causal measures are the risk difference, relative risk, and other relative risk, because they correspond to the number needed to treat, the attributable fraction, and causal power, respectively. It is therefore helpful to know that so long as we check that the relative risk and the other relative risk are both stronger in one stratum of the modifier versus another, that these three causal measures will all be stronger in that stratum. It does not hurt to know that the odds ratio will also be stronger in that stratum.

Following Brumback and Berg (2008) and Shannin and Brumback (2021), we also used a Monte-Carlo simulation study to investigate the chance that for randomly selected p_0, p_1, p_2, p_3 from the unit four-dimensional hypercube (e.g. randomly choose each proportion to be between 0 and 1, independently of one another), the various subsets of the four causal measures would agree (i.e. change together from one stratum to another). When RR_0 and RR_1 are on opposite sides of one, that is, when in one stratum the treatment is helpful and in the other it is harmful, then all four measures will automatically change together. Therefore, we repeated our simulation study to investigate the chance that the various subsets agree when we exclude the random selections of p_0, p_1, p_2, p_3 such that RR_0 and RR_1 are on opposite sides of one. The Venn diagrams in Figures 4.1 and 4.2 display the chances. We observe that when RR_0 and RR_1 are unconstrained, the four measures agree 83.3% of the time, whereas when we constrain RR_0 and RR_1 to be on the same side of one, the four measures only agree 66.7% of the time. In the unconstrained case, the chance that RR^*, RD, and OR agree with each other but not with RR is 5.7%, whereas in the constrained case, that chance increases to 11.4%. By symmetry, those probabilities are identical to the probabilities that RR, RD, and OR agree with each other but not with RR^*. For pairwise agreement, we see that the probability that any two measures agree with each other but not the other two is 2.6% in the unconstrained case and 5.3% in the constrained case. The probabilities shown in the Venn diagram do not add up to 100%, because, for example, the event that RR changes in the same direction as RD but not in the same direction as the other two measures is the same as the event that $RR*$ changes in the same direction as OR but not in the same direction as the other two. It would be awkward to arbitrarily display one of those two chances as zero.

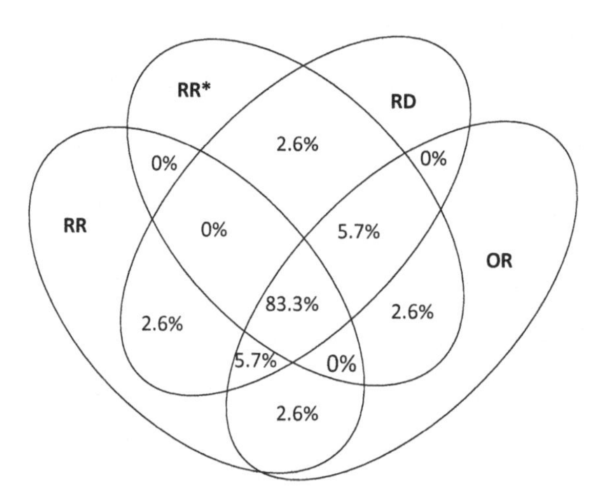

FIGURE 4.1: Monte-Carlo Study of Effect-Measure Modification: No Constraints

For a real example of agreement and disagreement, we compare the annual mortality rates in 2019 (`p19`) and 1984 (`p84`) across the 67 Florida counties. The 'treatment' T is the year and the modifier M is the county. The four measures `RD`, `RR`, `RRstar`, and `OR` and the underlying probabilities `p19` and `p84` are presented below for each county. When we considered all possible 2211 pairs of the 67 counties, `RR` and `RRstar` disagree for 106 (4.8%). We found that `RRstar` and `RD` always agree, as do `RR` and `OR`. Furthermore, `RD`

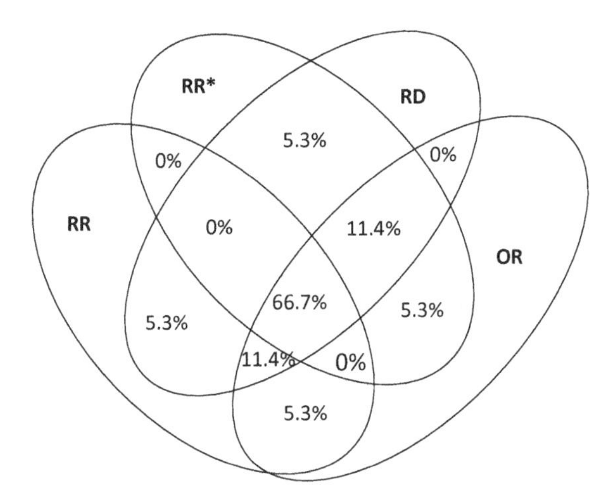

FIGURE 4.2: Monte-Carlo Study of Effect-Measure Modification: RR_0 and RR_1 on same side of one

and `RR` disagree on the same 106, as do `RRstar` and `OR`, and also `RD` and `OR`. Thus in this real example of agreement, we found that all four measures agree for 95.2% of the pairs. Focusing just on one pair, Alachua and Citrus, we see that for Alachua, $p_{19} = 0.00754$ and $p_{84} = 0.00611$ whereas for Citrus, $p_{19} = 0.0179$ and $p_{84} = 0.0146$. The risk difference is smaller (0.00143 versus 0.00332), but the relative risk is slightly larger (1.234 versus 1.227) for Alachua county versus Citrus.

	p19	p84	RD	RR	RRstar	OR
Alachua	0.0075417	0.0061103	1.4315e-03	1.23427	1.00144	1.23605
Baker	0.0093275	0.0061518	3.1757e-03	1.51623	1.00321	1.52109
Bay	0.0104113	0.0078181	2.5933e-03	1.33170	1.00262	1.33519
Bradford	0.0108944	0.0086146	2.2798e-03	1.26465	1.00230	1.26756
Brevard	0.0128368	0.0087631	4.0737e-03	1.46487	1.00413	1.47091
Broward	0.0079060	0.0112818	-3.3758e-03	0.70078	0.99660	0.69839
Calhoun	0.0110132	0.0120798	-1.0666e-03	0.91170	0.99892	0.91072
Charlotte	0.0150687	0.0146306	4.3811e-04	1.02994	1.00044	1.03040
Citrus	0.0179104	0.0145947	3.3157e-03	1.22719	1.00338	1.23133
Clay	0.0089264	0.0067060	2.2204e-03	1.33111	1.00224	1.33409
Collier	0.0095208	0.0099105	-3.8974e-04	0.96067	0.99961	0.96030
Columbia	0.0117540	0.0090461	2.7079e-03	1.29934	1.00274	1.30290
Miami.Dade	0.0070383	0.0096230	-2.5847e-03	0.73140	0.99740	0.72950
DeSoto	0.0098550	0.0123229	-2.4679e-03	0.79973	0.99751	0.79774
Dixie	0.0137442	0.0088385	4.9057e-03	1.55504	1.00497	1.56277
Duval	0.0090354	0.0086396	3.9586e-04	1.04582	1.00040	1.04624
Escambia	0.0115453	0.0077287	3.8167e-03	1.49383	1.00386	1.49960
Flagler	0.0137387	0.0106168	3.1219e-03	1.29406	1.00317	1.29815
Franklin	0.0130648	0.0131845	-1.1964e-04	0.99093	0.99988	0.99081
Gadsden	0.0100572	0.0091438	9.1338e-04	1.09989	1.00092	1.10091
Gilchrist	0.0110282	0.0103820	6.4616e-04	1.06224	1.00065	1.06293
Glades	0.0102306	0.0106896	-4.5898e-04	0.95706	0.99954	0.95662
Gulf	0.0112679	0.0104812	7.8672e-04	1.07506	1.00080	1.07592
Hamilton	0.0111584	0.0109796	1.7881e-04	1.01629	1.00018	1.01647
Hardee	0.0087877	0.0070835	1.7041e-03	1.24058	1.00172	1.24271
Hendry	0.0082067	0.0084103	-2.0354e-04	0.97580	0.99979	0.97560
Hernando	0.0150426	0.0138916	1.1510e-03	1.08286	1.00117	1.08412
Highlands	0.0161233	0.0142115	1.9117e-03	1.13452	1.00194	1.13672
Hillsborough	0.0076070	0.0084731	-8.6609e-04	0.89778	0.99913	0.89700
Holmes	0.0154813	0.0108567	4.6246e-03	1.42596	1.00470	1.43266
Indian.River	0.0141590	0.0117087	2.4503e-03	1.20927	1.00249	1.21228
Jackson	0.0126776	0.0110836	1.5940e-03	1.14382	1.00161	1.14567
Jefferson	0.0112519	0.0107469	5.0499e-04	1.04699	1.00051	1.04752
Lafayette	0.0094044	0.0090171	3.8726e-04	1.04295	1.00039	1.04335
Lake	0.0129154	0.0134468	-5.3139e-04	0.96048	0.99946	0.95997
Lee	0.0102569	0.0111401	-8.8323e-04	0.92072	0.99911	0.91989
Leon	0.0068921	0.0057752	1.1169e-03	1.19340	1.00112	1.19474
Levy	0.0133723	0.0117324	1.6399e-03	1.13978	1.00166	1.14167
Liberty	0.0076361	0.0075901	4.5953e-05	1.00605	1.00005	1.00610
Madison	0.0126453	0.0100045	2.6407e-03	1.26396	1.00267	1.26734
Manatee	0.0112263	0.0136637	-2.4373e-03	0.82162	0.99753	0.81959
Marion	0.0148589	0.0110740	3.7849e-03	1.34179	1.00384	1.34694
Martin	0.0124299	0.0120797	3.5019e-04	1.02899	1.00035	1.02935
Monroe	0.0090099	0.0086362	3.7367e-04	1.04327	1.00038	1.04366
Nassau	0.0109826	0.0068369	4.1457e-03	1.60636	1.00419	1.61310
Okaloosa	0.0093583	0.0059070	3.4513e-03	1.58428	1.00348	1.58980
Okeechobee	0.0116333	0.0106283	1.0050e-03	1.09456	1.00102	1.09567
Orange	0.0060606	0.0077381	-1.6775e-03	0.78321	0.99831	0.78189
Osceola	0.0068651	0.0094096	-2.5446e-03	0.72958	0.99744	0.72771
Palm.Beach	0.0101736	0.0117046	-1.5310e-03	0.86920	0.99845	0.86786
Pasco	0.0124266	0.0149346	-2.5079e-03	0.83207	0.99746	0.82996
Pinellas	0.0124332	0.0151043	-2.6711e-03	0.82315	0.99730	0.82093
Polk	0.0107133	0.0096795	1.0338e-03	1.10680	1.00105	1.10796
Putnam	0.0136553	0.0112318	2.4235e-03	1.21577	1.00246	1.21875
St..Johns	0.0084890	0.0099911	-1.5020e-03	0.84966	0.99849	0.84838
St..Lucie	0.0113436	0.0107243	6.1926e-04	1.05774	1.00063	1.05841

```
Santa.Rosa    0.0093565 0.0064457  2.9108e-03 1.45159 1.00294 1.45586
Sarasota      0.0140195 0.0144849 -4.6539e-04 0.96787 0.99953 0.96741
Seminole      0.0074729 0.0066577  8.1523e-04 1.12245 1.00082 1.12337
Sumter        0.0159520 0.0116139  4.3381e-03 1.37352 1.00441 1.37958
Suwannee      0.0127523 0.0110015  1.7507e-03 1.15914 1.00177 1.16119
Taylor        0.0115663 0.0108574  7.0893e-04 1.06529 1.00072 1.06606
Union         0.0157022 0.0075680  8.1342e-03 2.07483 1.00826 2.09197
Volusia       0.0137259 0.0131236  6.0231e-04 1.04590 1.00061 1.04653
Wakulla       0.0091924 0.0075145  1.6780e-03 1.22330 1.00169 1.22537
Walton        0.0096373 0.0094176  2.1966e-04 1.02332 1.00022 1.02355
Washington    0.0146763 0.0110770  3.5993e-03 1.32494 1.00365 1.32978
```

4.3 Causal Interaction

Causal interaction is a concept distinct from statistical interaction. As we have seen, statistical interaction concerns whether a causal effect differs across levels of an effect-modifier. With causal interaction, we are focused on two causes, and whether or not they synergize or antagonize in producing their effects. Let T_1 and T_2 denote the indicators for the two causes, and let Y denote the outcome. With two causes, each individual has four potential outcomes, denoted by $Y(t_1, t_2)$, for $t_1 = 0, 1$ and $t_2 = 0, 1$. For binary Y, the individual belongs to one of sixteen causal types, as shown in Table 4.4 (Hernan and Robins (2020)).

TABLE 4.4
Sixteen Causal Types for Two Causes

Type	$Y(0,0)$	$Y(0,1)$	$Y(1,0)$	$Y(1,1)$
1	0	0	0	0
2	0	0	0	1
3	0	0	1	0
4	0	0	1	1
5	0	1	0	0
6	0	1	0	1
7	0	1	1	0
8	0	1	1	1
9	1	0	0	0
10	1	0	0	1
11	1	0	1	0
12	1	0	1	1
13	1	1	0	0
14	1	1	0	1
15	1	1	1	0
16	1	1	1	1

If it is plausible that $T_1 = 1$ (relative to $T_1 = 0$) causes $Y = 1$ and $T_2 = 1$ causes $Y = 1$, it may also be possible that *monotonicity* holds for the causal types. That is, changing t_1 from 0 to 1 or t_2 from 0 to 1 cannot decrease $Y(t_1, t_2)$. This means that there are no individuals of causal types 3,5,7,9,10,11,12,13,14, and 15. The types remaining are 1,2,4,6,8, and 16. Type 2 is synergistic, because setting both T_1 and T_2 to one causes $Y = 1$ but either cause on its own is insufficient, as is the absence of both causes.

Consider, for example, the combined effects of T_1 as hormone replacement therapy (HRT) and T_2 as antioxidant vitamin supplements on death Y in postmenopausal women, investigated in the WAVE Trial, a four group randomized clinical trial reported in Waters et al. (2002). Death occured anytime during study follow-up, which was less than five years but presumably balanced across the four randomized treatment groups, which are $T_1 = T_2 = 1$; $T_1 = 1, T_2 = 0$; $T_1 = 0, T_2 = 1$; and $T_1 = T_2 = 0$. Data from Table 5 of Waters et al. (2002) are reproduced in Table 4.5 below

TABLE 4.5
Data from the WAVE Trial

T_1	T_2	$E(Y\|T_1, T_2)$	n
0	0	0.019	108
0	1	0.057	105
1	0	0.039	103
1	1	0.094	107

Ignoring statistical significance, the data suggest that combining T_1 and T_2 can be dangerous. To address this causally, we ask whether the data support existence of causal type 2 in the population, that is, people for whom HRT and antioxidant vitamins synergize to increase the risk of death. VanderWeele and Robins (2007) report that, under monotonicity of the causal types, a sufficient condition for the existence of causal type 2 is that

$$E(Y(1,1)) - E(Y(0,1)) - E(Y(1,0)) + E(Y(0,0)) > 0. \tag{4.1}$$

The proof is not difficult. We refer to Figure 4.3 for help representing probabilities with Venn diagrams. Let event A be that $Y(1,1) = 1$, event B be that $Y(0,1) = 1$, event C be that $Y(1,0) = 1$, and event D be that $Y(0,0) = 1$. For an event F, let $\bar{F}$ denote its opposite. Due to monotonicity, causal type 10 does not exist, and therefore the proportion of individuals of type 2 is

$$P(A \cap \bar{B} \cap \bar{C}) = P(A) - P(B \cap A) - P(C \cap A) + P(A \cap B \cap C).$$

Noting that monotonicity implies that D occurring requires that A, B, and C jointly occur, so that D is a subset of $A \cap B \cap C$, $P(A \cap B \cap C) \geq P(D)$. Furthermore, as $P(B \cap A) \leq P(B)$ and $P(C \cap A) \leq P(C)$, we have that

$$P(A \cap \bar{B} \cap \bar{C}) \geq P(A) - P(B) - P(C) + P(D). \tag{4.2}$$

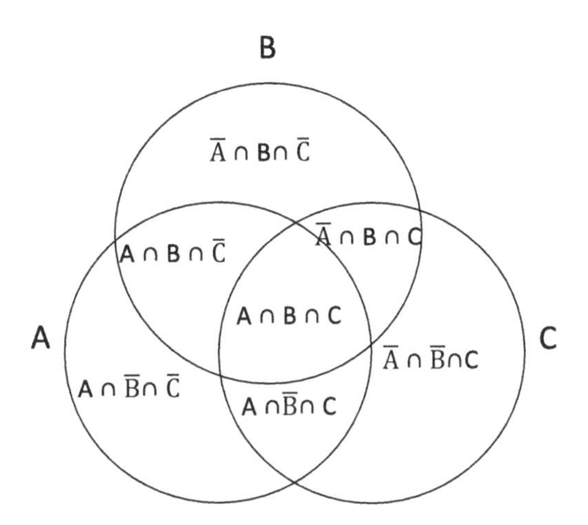

$$P(A) = P(A \cap \bar{B} \cap \bar{C}) + P(A \cap B \cap \bar{C}) + P(A \cap B \cap C) + P(A \cap \bar{B} \cap C)$$

FIGURE 4.3: Representing Probabilities with Venn Diagrams

Noting that $P(A) = E(Y(1,1))$, $P(B) = E(Y(0,1))$, $P(C) = E(Y(1,0))$, and $P(D) = E(Y(0,0))$, we can compute a lower bound for the proportion of individuals of type 2 below using (4.2) if we can compute the expected potential outcomes. For a randomized study such as the WAVE Trial, it is reasonable to assume

$$\{Y(0,0), Y(0,1), Y(1,0), Y(1,1)\} \amalg T_1, T_2, \tag{4.3}$$

and therefore we can estimate $E(Y(t_1, t_2))$ with an estimate of $E(Y|T_1 = t_1, T_2 = t_2)$.

Therefore, positive effect-modification of the risk difference is sufficient for synergism, assuming (4.3) and monotonicity. We use the data in Table 4.5 to estimate the lower bound at (4.2) by $0.094 - 0.039 - 0.057 + 0.019 = 0.017$. As the proportion of individuals of causal type 2 is greater than or equal to 0.017, it is greater than zero, and hence we estimate that there exist some such individuals in the population. Unfortunately, due to the small sample size of the study, our estimate is unlikely to be statistically significantly different from zero. and hence we cannot be certain. We leave the hypothesis test to the exercises.

Next we revisit the NCES data, letting T_1 be one minus `female`, T_2 be `selective`, and Y be `highmathsat`. We question whether admitting proportionally fewer women synergizes with admitting fewer students in raising math SAT scores of enrollees.

TABLE 4.6
Causal Interaction with NCES Data

T_1	T_2	$E(Y\|T_1, T_2)$
0	0	0.081
0	1	0.345
1	0	0.167
1	1	0.675

We can use the data of Table 4.6 to estimate the lower bound at (4.2) by $0.675 - 0.167 - 0.345 + 0.081 = 0.244$. Thus the data suggest that there is such synergy. However, in this example, we cannot rely on randomization to assure us that (4.3) holds. In fact, it is unlikely to hold, because selective schools, i.e. schools with $T_2 = 1$, are likely to have applicant pools with higher math SAT scores, in which case the independence would be violated. The causal interpretation is therefore poor, and we would not suggest schools admit fewer students and proportionally fewer women in order to raise the math SAT scores of enrollees.

4.4 Exercises

1. Using data from the 2007 Florida Behavorial Risk Factor Surveillance System Survey, Brumback et al. (2010) investigate modification by age group of the risk difference for the effect of having a disability on having a cost barrier to health care. They also present the estimated risks. All estimates are presented in Brumback et al. (2010) Table 4, and they have been adjusted for confounding by race/ethnicity, income, education, and gender. The risk of a cost barrier for persons with disability are 0.383 and 0.072 for persons 18–29 years of age and greater than 65 years of age, respectively, whereas for persons without a disability they are 0.225 and 0.040 for those two age groups. Compute the RD, RR, RR*, and OR for each age group. Is there a clear answer to determining which age group has a stronger effect of disability on having a cost barrier to health care? Suppose we can interpret the results causally. Suppose policymakers have limited resources and ask which age group would benefit most from an intervention to provide assistance to persons with disability in terms of dissolving the cost barrier. What would you tell them?
2. The `sepsis` binary dataset is taken from a prospective study conducted by the University of Florida Sepsis and Critical Illness Research Center; see Loftus et al. (2017) for the study design. All patients were septic at study entry. The variables are `gt65` indicating greater than 65 years of age, `shock` indicating presence of septic shock at study entry, and `zubrod45` indicating a Zubrod score of 4 or 5 one year after study entry, where a score of 4 means confined to bed at all times and a score of 5 means death. Use the data to compute the RD, RR, RR*, and OR for each age group to investigate the effect of septic shock on a Zubrod score of 4 or 5 one year out. Also compute 95% confidence intervals for each measure and for the contrast of each measure across the two age groups. Is there a clear answer to determining which age group has a stronger effect?

```
> head(sepsis)
   gt65 shock zubrod45
1     1     1        1
5     0     0        0
7     0     0        0
8     1     1        1
9     1     0        0
10    1     1        1
```

3. For the `brfss` data, determine whether `whitenh` modifies the effect of `gthsedu` on `insured` in respondents less than 65 years of age. What assumptions would be necessary for a causal interpretation? Are these likely to hold? If a causal interpretation were possible, how would you interpret your results?

4. For the WAVE Trial, with data presented in Table 4.5, determine if the apparent causal interaction between hormone replacement therapy and antioxidant vitamin supplements on death is statistically significant, assuming monotonicity of the causal types.

5. For the `gss` data, construct a complete-case dataset with `attend`, `gthsedu`, and `female`, and use it to investigate the possibility of a causal synergy of `gthsedu` and `female` on `attend`. Some researchers, e.g. Holland (1986), believe that `female` cannot be a cause but can only be an attribute. We take the position that `female` is standing in for social-cultural forces that tend to influence women more than men but that could influence either. What additional assumptions would be necessary for a causal interpretation? Are these likely to hold?

6. Suppose that

$$\{Y(0,0), Y(0,1), Y(1,0), Y(1,1)\} \amalg T_1, T_2$$

does not hold but that

$$\{Y(0,0), Y(0,1), Y(1,0), Y(1,1)\} \amalg T_1, T_2 | H \tag{4.4}$$

does hold. Show how to make use of assumption (4.4) to modify the sufficient condition for synergy, assuming monotonicity.

5

Causal Directed Acyclic Graphs

5.1 Theory

Causal directed acyclic graphs provide a convenient and efficient way to represent statistical and causal dependence among a collection of variables. We rely heavily on Pearl (1995) and Greenland et al. (1999) for our overview. Given the collection $X_1, \ldots, X_p$, repeated application of the multiplication rule allows us to write their joint distribution as

$$P(X_1, \ldots, X_p) = \Pi_{j=1}^{p} P(X_j | X_1, \ldots, X_{j-1}), \tag{5.1}$$

where X_0 is the empty set. The variables $(X_1, \ldots, X_{j-1})$ are termed the *predecessors* of X_j. Suppose $P(X_j|X_1, \ldots, X_{j-1}) = P(X_j|pa_j)$, where pa_j are a select group of the predecessors of X_j, termed the *parents* of X_j, and suppose also that pa_j is the minimal such set. For X_1, let pa_j be the empty set. Then we can write the joint distribution (5.1) as

$$P(X_1, \ldots, X_p) = \Pi_{j=1}^{p} P(X_j | pa_j). \tag{5.2}$$

Notice that (5.2) encodes conditional independences not specified by (5.1). We can draw a directed acyclic graph (DAG) that also encodes these conditional independences.

To do so, start with earlier variables and draw an arrow from X_h to X_j $(h < j)$ unless $X_j \amalg X_h | pa_j$. For example,

$$P(X_1, X_2, X_3, X_4) = P(X_4|X_2, X_1)P(X_3|X_2)P(X_2|X_1)P(X_1) \tag{5.3}$$

corresponds to the DAG of Figure 5.1.

We have constructed a graph encoding the conditional independences that are obvious from (5.3), which are $X_4 \amalg X_3 | X_2, X_1$ and $X_3 \amalg X_1 | X_2$. A useful theorem states that we can use the DAG to determine independences implied by the obvious ones that are not themselves obvious from (5.3). For our example, we also have that $X_4 \amalg X_3 | X_2$ and $X_3 \amalg X_1 | X_2, X_4$. These additional independences are difficult to determine directly from the factorization (5.3). One needs to apply a lot of algebraic manipulation. That is why the theorem

DOI: 10.1201/9781003146674-5

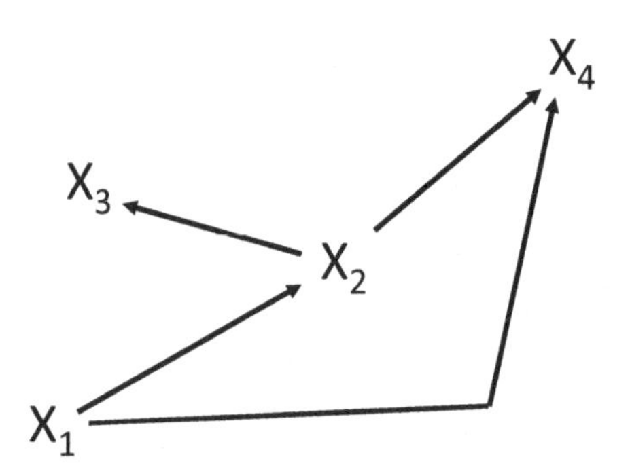

FIGURE 5.1: Example of a Directed Acyclic Graph

allowing us to determine the conditional independences from the graphs is useful.

To prepare us for the theorem, we need to learn *d-separation*. Consider 3 disjoint sets of variables A, B, and C, represented as nodes on a DAG. A *path* is a sequence of consecutive edges (of any directionality) in the graph. A path is said to be d-separated, or blocked, by a set of variables C if and only if the path (i) contains a *chain* as in Figure 5.2a, such that the middle variable Z is in C, or a *fork* as in Figure 5.2b, such that the middle variable Z is in C, or (ii) contains an inverted fork, or *collider*, as in Figure 5.2c such that the middle variable Z is not in C and such that no descendant of a collider is in C (i.e. in 5.2d, Z cannot be in C, and neither can be W.) Then the set C is said to *d-separate* A from B if and only if C blocks every path from a variable in A to a variable in B.

Based on this definition of d-separation, the useful theorem can be stated as follows.

> *Theorem:* If A and B are d-separated by C in a DAG, then $A \amalg B|C$. Conversely, if A and B are not d-separated by C in the DAG, then A and B are dependent conditional on C unless the dependences represented by the arrows exactly cancel.

The requirement that the dependences represented by the arrows do not exactly cancel is known as *faithfulness* (Zhang and Spirtes (2008)). In Chapter 6, we present an analysis of the General Social Survey data in which the faithfulness assumption is violated. For help in applying this theorem, we will review the four basic graphical structures shown in Figure 5.2 and the independences implied by each.

- In Figure 5.2a, Z is an *intermediate variable.* The only independence implied by this structure is $X_1 \amalg X_2|Z$. It is NOT true that $X_1 \amalg X_2$. In the context of the What-If? study, suppose X_1 indicates being randomized to treatment

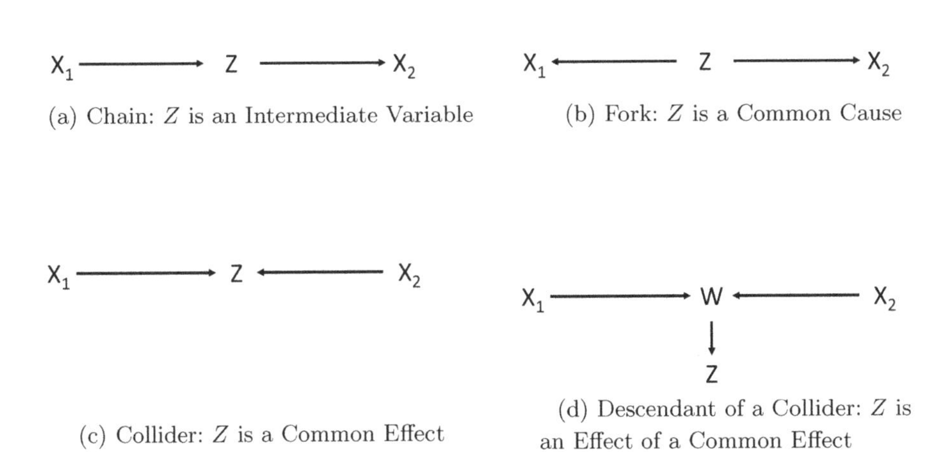

(a) Chain: Z is an Intermediate Variable

(b) Fork: Z is a Common Cause

(c) Collider: Z is a Common Effect

(d) Descendant of a Collider: Z is an Effect of a Common Effect

FIGURE 5.2: Graphical Structures

with naltrexone, Z indicates reducing drinking, and X_2 indicates subsequent health. It is reasonable to assume that naltrexone does not directly cause subsequent health, but indirectly can cause it via reduced drinking. Thus if our investigation is restricted to people who have reduced their drinking and have $Z = 1$, there would no longer be an association between having taken naltrexone and subsequent health; therefore, $X_1 \perp\!\!\!\perp X_2 | Z = 1$, and similarly for $Z = 0$. Because naltrexone can cause subsequent health indirectly, we have $X_1 \not\perp\!\!\!\perp X_2$.

- In Figure 5.2b, Z is a *common cause.* The only independence implied by this structure is $X_1 \perp\!\!\!\perp X_2 | Z$. It is NOT true that $X_1 \perp\!\!\!\perp X_2$. Suppose Z indicates healthy behaviors, which can cause adherence to HIV medication $X_1 = 1$ as well as healthy diet $X_2 = 1$. Supposing we know someone has adherence to HIV medication, we would predict he or she had healthy behaviors, and therefore we would also predict healthy diet; thus, $X_1 \not\perp\!\!\!\perp X_2$. However, suppose we look only at people with healthy behaviors $Z = 1$; then we would expect everyone to be adherent to HIV medication and have a healthy diet. With X_1 and X_2 constant, there is no statistical association; that is, knowledge of adherence to HIV medication does not effect prediction of healthy diet, because we know that everyone has a healthy diet. Therefore, $X_1 \perp\!\!\!\perp X_2 | Z = 1$. Analogously, for $Z = 0$, we would expect everyone to fail to adhere to their medication and not have a healthy diet, so that $X_1 \perp\!\!\!\perp X_2 | Z = 0$.

- In Figure 5.2c, Z is a *common effect.* The only independence implied by this structure is $X_1 \amalg X_2$. It is NOT true that $X_1 \amalg X_2|Z$. Suppose X_1 indicates a selective university and X_2 indicates admitting proportionally more men, and Z indicates a high average math SAT score of enrollees. Knowing a university is selective does not give us information on whether or not it admits proportionally more men, so that $X_1 \amalg X_2$. However, If we look only at universities with high average math SAT scores, knowing that the university is not selective would induce us to predict that it admits proportionally more men, so that $X_1 \not\amalg X_2|Z$.

- In Figure 5.2d, Z is an *effect of a common effect.* The independences implied by this structure are $X_1 \amalg X_2$, $Z \amalg X_1|W$, and $Z \amalg X_2|W$. It is NOT true that $X_1 \amalg X_2|W$ or that $X_1 \amalg X_2|Z$. This is the trickiest of the four types of structures. Now let X_1 indiciate a selective university, X_2 indicate admitting proportionally more men, W indicate high average math SAT score of enrollees, and Z indicate high average math GRE score of graduates. $X_1 \amalg X_2$ for the same reasons as in the previous example, but $X_1 \not\amalg X_2|Z$ because knowing the school has high average math GRE scores of graduates induces us to predict it has high math SAT scores of enrollees, and therefore X_1 and X_2 are associated again given Z for the same reasons as in the previous example.

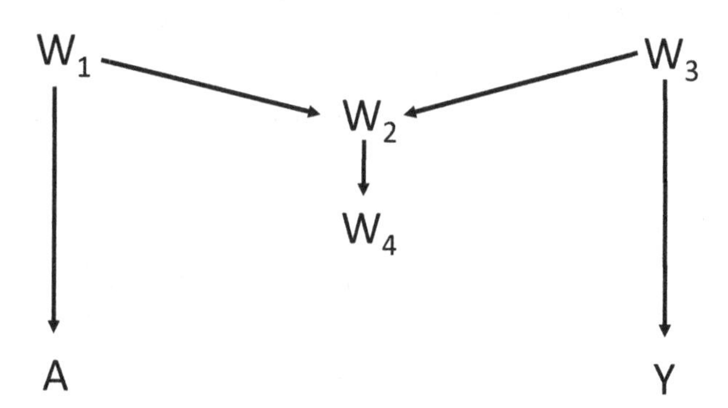

FIGURE 5.3: Causal DAG With Collider

To practice applying the theorem, consider the DAG of Figure 5.3. Some of the independences that the DAG implies are

$$
\begin{aligned}
A &\perp\!\!\!\perp Y \\
A &\perp\!\!\!\perp W_3 \\
A &\perp\!\!\!\perp Y \mid W_3 \\
A &\perp\!\!\!\perp W_3 \mid Y \\
A &\perp\!\!\!\perp Y \mid W_1, W_2 \\
A &\perp\!\!\!\perp Y \mid W_4, W_3 \\
W_1 &\perp\!\!\!\perp W_3 \\
W_1 &\perp\!\!\!\perp Y \\
W_1 &\perp\!\!\!\perp Y \mid A \\
W_4 &\perp\!\!\!\perp Y \mid W_2.
\end{aligned}
$$

Some of the dependences that the DAG implies, assuming faithfulness, are

$$
\begin{aligned}
A &\not\!\perp\!\!\!\perp W_1 \\
A &\not\!\perp\!\!\!\perp W_2 \\
A &\not\!\perp\!\!\!\perp Y \mid W_2 \\
A &\not\!\perp\!\!\!\perp Y \mid W_4 \\
A &\not\!\perp\!\!\!\perp W_1 \mid W_2, W_4 \\
W_1 &\not\!\perp\!\!\!\perp W_2 \\
W_1 &\not\!\perp\!\!\!\perp W_4 \\
W_1 &\not\!\perp\!\!\!\perp W_3 \mid W_2 \\
W_1 &\not\!\perp\!\!\!\perp W_3 \mid W_2, W_4 \\
W_3 &\not\!\perp\!\!\!\perp Y.
\end{aligned}
$$

One might wonder if the DAG of Figure 5.3 could correspond to any real example. Indeed, it is quite difficult to come up with real examples in which the missing arrows are plausible. Consider the following attempt to characterize a randomized trial in a population of post-menopausal women having difficulty sleeping. Let W_1 be randomization to a certain sedative, which sometimes causes headache disorder, A, but can induce more than 6 hours of sleep per night, W_2. Let W_3 be randomization to hormone repaclement therapy (HRT) supplementation, which can also aid in sleep (W_2) due to reduction of hot flashes, but carries a risk of increased blood pressure, Y. Finally, let W_4 indicate increased productivity at work. One might wonder if the headache disorder A and the increased blood pressure Y are related; the DAG says they are not. The difficulty with this example, as with many others, in satifying the assumptions of the DAG, is that sleeping less than 6 hours per night might induce a headache disorder, and a headache disorder might reduce productivity at work, and less sleep as well as unproductive work days might increase blood pressure. It is often too hard to justify the missing arrows.

The previous theorem is most useful because it implies another theorem which helps us to identify causal effects by identifying a sufficient set of confounders. Specifically,

> *Backdoor Theorem:* Given a DAG containing two variables A and Y as well as a set of variables C excluding A and Y that does not contain descendants of A nor Y, the set C is sufficient to adjust for confounding of the effect of A on Y if and only if there exists *no unblocked backdoor path* from A to Y. That is,
>
> $$\{Y(a)\}_{a \in \mathcal{A}} \amalg A|C,$$
>
> where $\{Y(a)\}$ is the set of potential outcomes to all possible values $a \in \mathcal{A}$ of A.

Again considering our example DAG of Figure 5.3, the empty set (that is, no variables in C) is sufficient to adjust for confounding of the effect of A on Y, but so is the set W_1 or the set (W_1, W_2) or (W_3, W_4). It is noteworthy that the variable W_2 is insufficient by itself to adjust for confounding of the effect of A on Y. This insufficiency proves that a traditional definition of confounder as a variable that is associated with A and Y is inadequate, because W_2 is such a variable and because the DAG implies that adjusting for it is worse than not adjusting for any variables. Adjusting for W_2 or W_4 without also adjusting for W_1 or W_3 can cause a bias referred to as *collider bias.* Referring back to our definition of *true confounder* in Chapter 1, i.e. a variable that influences the exposure and that also influences the outcome via a directed path that does not include the exposure, we see that W_2 does not satisfy the criteria, as it influences neither A nor Y. In fact, none of the variables in the DAG of Figure 5.3 satisfies the criteria, and therefore, there is no true confounder and hence no confounding of the effect of A on Y. We learn from this example that a sufficient set of confounders need not contain a true confounder. However, if there is no true confounder in a DAG, then the empty set is a sufficient set of confounders. As mentioned in Chapter 1, in some cases not all of the true confounders are required to form a *sufficient set of true confounders,* which is simply a sufficient set of confounders that are all true confounders. In the next section and in the exercises we consider several more examples.

5.2 Examples

Figure 5.4 presents a causal DAG showing the intermediate variable M mediating the effect of A on Y. Suppose the arrow from A directly to Y is missing. In that case, it is tempting to think that conditioning on M would block the mediation from A to Y, so that $A \amalg Y|M$; however, because M is a collider on the path from A to H to Y, conditioning on it would open up that path, so that in fact, $A \not\!\perp\!\!\!\perp Y|M$. The confounding by H of the effect of M on Y needs to be accounted for in mediation analyses; Chapter 12 explains how.

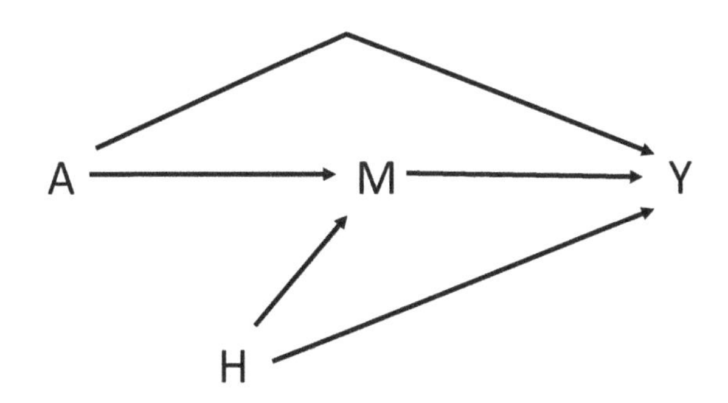

FIGURE 5.4: Causal DAG With Intermediate Variable

Figure 5.5 illustrates an unmeasured true confounder U of the effect of A on Y that can be handled by conditioning on the measured variable Z, even if Z occurs *after* A. By the Backdoor Theorem, we have that $\{Y(a)\}_{a\in\mathcal{A}} \amalg A|Z$.

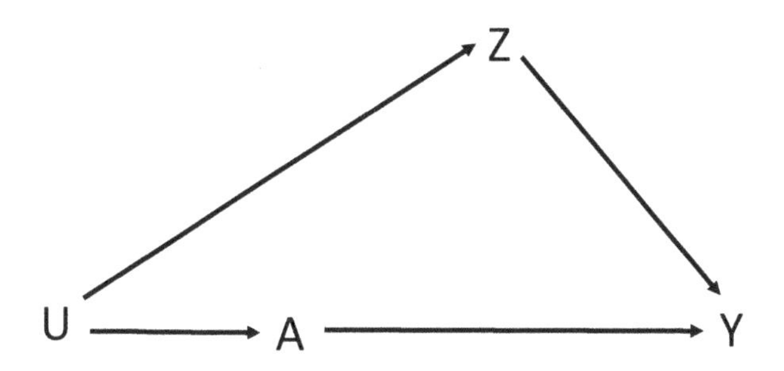

FIGURE 5.5: A Confounder May Occur After the Exposure

The collection of potential outcomes $\{Y(a)\}_{a\in\mathcal{A}}$ can be viewed as the ultimate confounder, even when it is not a true confounder. The unmeasured confounder $U = \{Y(a)\}_{a\in\mathcal{A}}$ will always block all backdoor paths from A to Y, because Y is a deterministic function of A and $\{Y(a)\}_{a\in\mathcal{A}}$; that is, $Y = Y(A)$, or, for binary Y, $Y = AY(1) + (1 - A)Y(0)$. This is represented by the causal DAG of Figure 5.6. The `R` code `sim1.r` below illustrates a possible data generating mechanism.

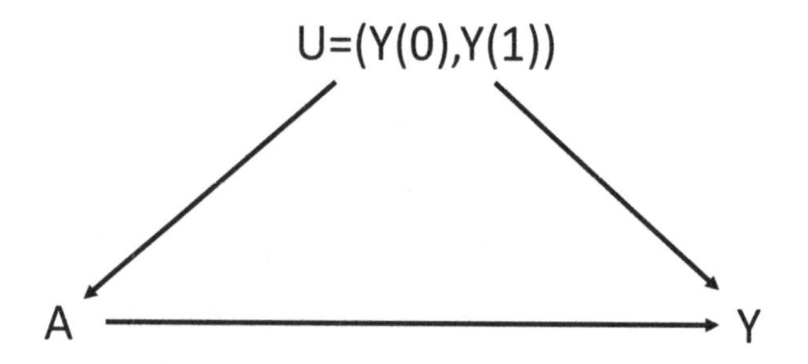

FIGURE 5.6: One Way to Generate Confounding

```
sim1.r <- function ()
{
  set.seed(111)
  #Generate the potential outcomes first
  probY0 <- .42
  probY1 <- .62
  Y0 <- rbinom(n = 1000, size = 1, prob = probY0)
  Y1 <- rbinom(n = 1000, size = 1, prob = probY1)
  #Let the treatment depend on the potential outcomes
  probA <- (1 - Y0) * (1 - Y1) * .6307 +
    (1 - Y0) * Y1 * .4867 + Y0 * (1 - Y1) * .4699 +
    Y0 * Y1 * .4263
  A <- rbinom(n = 1000, size = 1, prob = probA)
  #Y must depend on A, Y1, and Y0 in this way
  Y <- A * Y1 + (1 - A) * Y0
  out <- data.frame(cbind(A, Y0, Y1, Y))
  out
}
```

Table 5.1 presents the probabilities corresponding to simulation `sim1.r`; these can be computed using the laws of probability and the multiplication rule. Take the first row, for example, the probability can be computed as $P(A = 0|Y(0) = 0, Y(1) = 0)P(Y(1) = 0|Y(0) = 0)P(Y(0) = 0)$ by the multiplication rule. From the simulation, we see that when $Y(0) = Y(1) = 0$, `probA`$= 0.6307$, which means that $P(A = 0|Y(0) = 0, Y(1) = 0) = 0.3693$. The probability $Y(0) = 0$ is 0.58, and the probability $Y(1) = 0$ given $Y(0) = 0$ is 0.38. Therefore, the probability of the data in the first row is $0.3693 * 0.58 * 0.38 = 0.0814$.

TABLE 5.1
Simulation Probabilities for `sim1.r`

A	$Y(0)$	$Y(1)$	Y	prob
0	0	0	0	0.0814
0	0	1	0	0.1846
0	1	0	1	0.0846
0	1	1	1	0.1494
1	0	0	0	0.1390
1	0	1	1	0.1750
1	1	0	0	0.0750
1	1	1	1	0.1110

The causal DAG of Figure 5.7 shows another way to generate confounding of the effect of A on Y; this time, there is no reference to $\{Y(a)\}_{a \in \mathcal{A}}$. The R code `sim2.r` illustrates a data generating mechanism that yields the same joint distribution of A and Y as `sim1.r`.

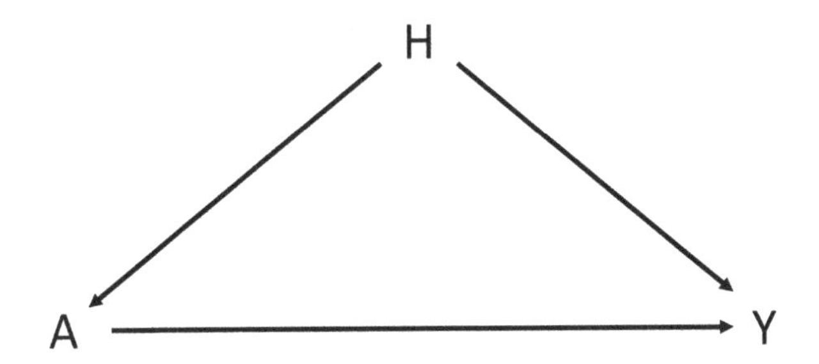

FIGURE 5.7: Another Way to Generate Confounding

```
sim2.r <- function ()
{
  set.seed(222)
  # Generate the confounder H first
  H <- rbinom(n = 1000, size = 1, prob = .4)
  # Let the treatment depend on the confounder
  probA <- H * .8 + (1 - H) * .3
  A <- rbinom(n = 1000, size = 1, prob = probA)
  # Let the outcome depend on the treatment and the confounder
  probY <- A * (H * .5 + (1 - H) * .7) + (1 - A) * (H * .3 +
      (1 - H) * .5)
  Y <- rbinom(n = 1000, size = 1, prob = probY)
  out <- data.frame(cbind(H, A, Y))
  out
}
```

Table 5.2 presents the probabilities corresponding to simulation `sim2.r`, which can also be computed using the laws of probability and the multiplication rule. For the first row, $P(H = 0) = 0.6$, and $P(A = 1|H = 0) = 0.3$, so that $P(A = 0|H = 0) = 0.7$, and $P(Y = 1|A = 0, H = 0) = 0.5$, so that $P(Y = 0|A = 0, H = 0) = 0.5$. Therefore, the probability of the data in the first row is $P(Y = 0|A = 0, H = 0)P(A = 0|H = 0)P(H = 0) = 0.5 * 0.7 * 0.6 = 0.21$.

TABLE 5.2
Simulation Probabilities for `sim2.r`

A	H	Y	prob
0	0	0	0.21
0	0	1	0.21
0	1	0	0.056
0	1	1	0.024
1	0	0	0.054
1	0	1	0.126
1	1	0	0.16
1	1	1	0.16

For both `sim1.r` and `sim2.r`, and thus for both Tables 5.1 and 5.2, the joint distribution of A and Y is given by Table 5.3. For example, for the first row, the probability $P(A = 0, Y = 0)$ can be calculated by the laws of probability as $P(A = 0, Y(0) = 0, Y(1) = 0, Y = 0) + P(A = 0, Y(0) = 0, Y(1) = 1, Y = 0) = 0.0814 + 0.1846 = 0.266$, and it also equals $P(A = 0, H = 0, Y = 0) + P(A = 0, H = 1, Y = 0) = 0.21 + 0.056 = 0.266$.

TABLE 5.3
Joint Distribution of A and Y for `sim1.r` and `sim2.r`

A	Y	Prob
0	0	0.266
0	1	0.234
1	0	0.214
1	1	0.286

Furthermore, as we can easily compute for `sim1.r`, $E(Y(1)) - E(Y(0)) =$

$0.62 - 0.42 = 0.2$. Using the methods of Chapter 6, we will be able to compute that `sim2.r` also yields $E(Y(1)) - E(Y(0)) = 0.2$. One might wonder if knowledge of (A, H, Y) is equivalent to knowledge of $(A, Y(0), Y(1), Y)$. Given (A, Y), we cannot recover both $Y(0)$ and $Y(1)$. For example, when $A = 1$ and $Y = 1$, we know that $Y(1) = 1$ because it equals Y, but from Table 5.1 we know only that $P(Y(0) = 1|A = 1, Y = 1) = 0.110/0.286 = 0.385$; it is more likely that $Y(0) = 0$, but it also may transpire that $Y(0) = 1$. Does the additional knowledge of H help to identify $Y(0)$? We leave that as a question for the reader.

The causal DAG of Figure 5.8 relates the causal DAGs of Figures 5.6 and 5.7. We see that even when the data generating mechanism makes no reference to the potential outcomes, as in `sim2.r` and Figure 5.7, we can redraw the DAG to include the potential outcomes behind the scenes.

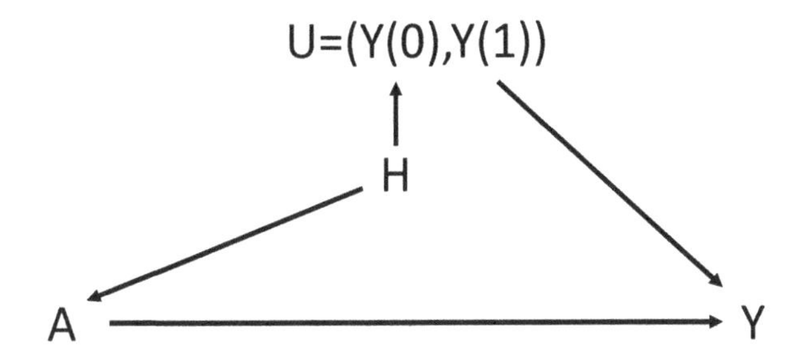

FIGURE 5.8: Potential Outcomes Behind the Scenes

We will say that a directed path from one variable to another is a mediated directed path if it passes through a third variable included in the DAG. Otherwise it is an unmediated directed path. When we redraw a causal DAG to include the potential outcomes behind the scenes, we must substitute each unmediated directed path from a variable to the outcome Y with a mediated directed path that passes through the potential outcomes, except that we do not substitute for any paths that are included as part of directed paths from the treatment A to the outcome Y. We must do this because Y is a deterministic function of the potential outcomes and A. The exercises will provide practice with this. Notice that we have not altered the dependence structure of the original variables in the DAG. With this simplified casual DAG, the proof of the Backdoor Theorem is easy. We need only use the d-separation theorem to find a vector of variables H satisfying $\{Y(0), Y(1)\} \amalg A|H$, and then that H is automatically a sufficient confounder for the effect of A on Y.

Although we have managed to insert potential outcomes into a DAG without altering its independence structure, we still need to verify that we have not altered the joint distribution of (V, A, Y), where here V represents all of the variables of the original DAG except for A and Y. That is, we need to make sure that there exists a joint distribution for $(V, A, Y(0), Y(1))$ that is compatible with (V, A, Y). Denote the joint distribution of (V, A, Y) as $P^*(V, A, Y)$. Without loss of generality, suppose $A = 1$. Then $P(V, A = 1, Y(0), Y(1)) = P(V, A = 1, Y, Y(0)) = P(Y(0)|V, A = 1, Y)P(V, A = 1, Y)$. We can therefore let $P(V, A = 1, Y) = P^*(V, A = 1, Y)$ and carefully choose a distribution for $P(Y(0)|V, A = 1, Y)$ that preserves the relationships between the components of $(Y(0), V, A, Y)$ implied by the consistency assumption and the causal DAG. For example, supposing $V = H$ is a sufficient confounder for the effect of A on Y, we have that $E_{Y|H,A=1}E(Y(0)|H, A = 1, Y)$ must equal $E(Y|A = 0, H)$; therefore, we cannot choose the distribution $P(Y(0)|V, A = 1, Y)$ arbitrarily. In this way, we can select a distribution for $(V, A = 1, Y(0), Y(1))$ that is compatible with $P^*(V, A = 1, Y)$. As the same argument holds for any possible value of A, we have verified our claim.

Next we present the simulation we used to construct `sim1.r` and `sim2.r`. In `simall.r`, we used the multiplication rule to specify a full joint distribution for $(H, A, Y(0), Y(1), Y)$; i.e., the simulation corresponds to the causal DAG of Figure 5.8. From this, we calculated the conditional probabilities used with the multiplication rule in `sim1.r` and `sim2.r` to generate the joint distributions for $(Y(0), Y(1), A, Y)$ and (H, A, Y) that are compatible with that for $(H, A, Y(0), Y(1), A, Y)$. Note that without simulating the full joint distribution first, it would be quite difficult to construct one that is compatible with both the distributions simulated by `sim1.r` and `sim2.r` such that all causal effects implied by the two distributions agreed.

```
simall.r
function ()
{
set.seed(888)
# Generate the observed confounder
H<-rbinom(n=1000,size=1,prob=.4)
# Generate the treatment
probA<-H*.8+(1-H)*.3
A<-rbinom(n=1000,size=1,prob=probA)
# Generate the potential outcomes dependent on the observed confounder
probY0<-H*.3 + (1-H)*.5
probY1<-H*.5 + (1-H)*.7
Y0<-rbinom(n=1000,size=1,prob=probY0)
Y1<-rbinom(n=1000,size=1,prob=probY1)
# Generate Y as a function of A, Y0, and Y1
Y<-A*Y1 + (1-A)*Y0
out<-data.frame(cbind(H,A,Y0,Y1,Y))
out
}
```

We conclude with the causal DAG used to generate data for the Double What-If? Study, introduced in Chapter 1 and reproduced in Figure 5.9. Several questions could be answered using data from the study. For example, one might ask what is the *intent-to-treat* effect of naltrexone ($T = 1$) on end of study viral load (VL_1)? The intent-to-treat effect, discussed at more length in Chapter 9, simply compares the outcome across the two randomized groups, $T = 1$ and $T = 0$, without concern whether participants assigned to naltrexone actually took naltrexone. The causal DAG shows that naltrexone was randomized, as there are no arrows into T. Therefore, we do not need to worry about confounding of the effect of T on VL_1. However, if we were interested in the effect of reducing drinking ($A = 1$) on VL_1, we need to worry about the true confounder U, which has an arrow into A and a directed path from U to VL_1 mediated by AD_0. The Backdoor Theorem tells us that AD_0 is sufficient to adjust for confounding of the effect of A on VL_1; letting $VL_1(0)$ and $VL_1(1)$ be the potential outcomes to $A = 0$ and $A = 1$, the Backdoor Theorem tells us that $\{VL_1(0), VL_1(1)\} \amalg A|AD_0$. The Backdoor Theorem also tells us that VL_0 is insufficient to adjust for that confounding; that is, $\{VL_1(0), VL_1(1)\} \not\amalg A|VL_0$. In the next chapter, we will learn how to use standardization to adjust for the effect of A on VL_1. In chapter 7, we will compare standardization to a difference-in-differences approach to adjust for confounding, which makes use of VL_0 rather than AD_0, and is valid under a different assumption. We introduce three simple difference-in-difference approaches to adjusting for confounding, and the one that is valid for the Double What-If? Study relies on the assumption that

$$E(VL_1(0)|A = 1) - E(VL_1(0)|A = 0) = E(VL_0|A = 1) - E(VL_0|A = 0),$$

called *additive equi-confounding* (Hernan and Robins (2020)). Thus, the Backdoor Theorem gives us one way to adjust for confounding, but there are other ways to consider as well. Yet another method uses T as an *instrumental variable* to adjust for confounding of the effect of A on VL_1, the subject of Chapter 9. In Chapter 12, we will also use the causal DAG of Figure 5.9 to address the question: Is the effect of assignment to naltrexone $T = 1$ on viral load at end of study VL_1 mediated by reducing drinking $A = 1$? This mediation analysis can use AD_0 to account for the confounding of the effect of A on VL_1, because the DAG has a structure similar to Figure 5.4, despite the extra variables.

Although a correct causal DAG is very useful for mapping out an appropriate statistical analysis plan, correctly constructing a causal DAG for a scientific study is generally quite difficult. Readers may find the `R` packages `ggdag` and `dagitty` helpful for producing DAGs for publication, but in most cases, the causal DAG is likely to be misspecified. For example, in Figure 5.3, an arrow from W_2 or W_4 to A or Y would change the confounding structure drammatically. Rather than having no confounding, there would be a true confounder and we would need to adjust our analysis. Supposing the arrow was from W_4 to Y, then W_1 would be a true confounder and also sufficient for confounding adjustment. Supposing the arrow was from W_4 to A, then W_3 would

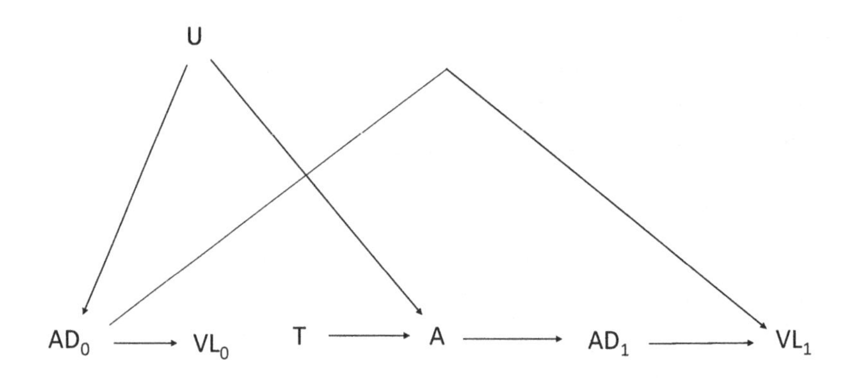

FIGURE 5.9: Causal DAG for Double What-If? Study

be a true confounder and also sufficient for confounding adjustment. Supposing arrows both from W_4 to A and from W_4 to Y, then (W_1, W_2, W_3, W_4) would be a true confounder and also sufficient for confounding adjustment. It may be the case that causal DAGs are more useful for destructive purposes than constructive purposes. Given a putative causal DAG and corresponding statistical analysis plan, a reviewer can argue that missing arrows should instead be present, or that other variables need to be inserted into the DAG. The reviewer's DAG will often show the original statistical analysis plan to be biased. However, even if the impact is to induce caution rather than to offer validation, causal DAGs can play an important role in causal inference.

5.3 Exercises

1. The hypothetical Get-in-Shape Study (GISS) randomized participants in equal numbers to an exercise program, a healthy diet, neither, or both. Measured outcomes were weight loss, increased strength, and increased nutrient intake. Match the variables of the GISS to the causal DAG of Figure 5.10. Given that DAG, how would you analyze the effect of increased nutrient intake on increased strength? Suppose you conducted that analysis, accompanied by the DAG. How might a reviewer critique your results?

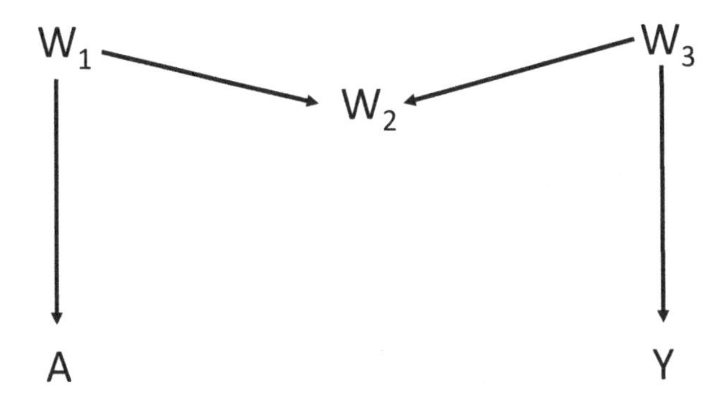

FIGURE 5.10: Causal DAG for Exercise 1

2. Consider the causal DAG of Figure 5.11. Which variables are true confounders for the effect of A on Y? Is it necessary to measure all of them, or is there a reduced set of true confounders that are sufficient for the effect of A on Y? Is the set (H_2, H_4) a sufficient set of confounders, a sufficient set of true confounders, or neither? Redraw the DAG to include the potential outcomes behind the scenes, and verify that they form a sufficient set of confounders. Are the potential outcomes true confounders?

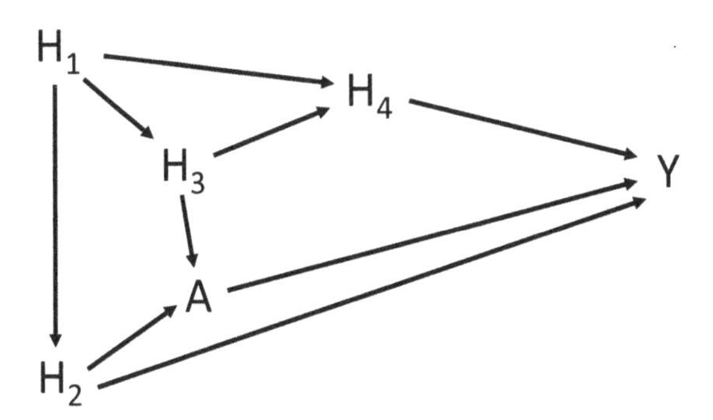

FIGURE 5.11: Causal DAG for Exercise 2

3. Draw a causal DAG corresponding to the R code of simex3.r below. List the true confounders. List the smallest sufficient set of true confounders. Redraw the DAG to include the potential outcomes behind the scenes, and verify that they form a sufficient set of confounders. Are the potential outcomes true confounders?

```
simex3.r <- function ()
 {
   set.seed(1234)
   probH1 <- 0.1
   H1 <- rbinom(n = 1000, size = 1, prob = probH1)
   probH2 <- 0.1 + 0.1 * H1
   H2 <- rbinom(n = 1000, size = 1, prob = probH2)
   probH3 <- 0.2 + 0.2 * H2
   H3 <- rbinom(n = 1000, size = 1, prob = probH3)
   probH4 <- 0.5
   H4 <- rbinom(n = 1000, size = 1, prob = probH4)
   probA <- 0.1 + 0.1 * H2 + 0.1 * H3 + 0.1 * H4
   A <- rbinom(n = 1000, size = 1, prob = probA)
   probY <- 0.1 + 0.2 * H1 + 0.2 * H2 + 0.3 * A
   Y <- rbinom(n = 1000, size = 1, prob = probY)
   out <- cbind(H1, H2, H3, H4, A, Y)
   data.frame(out)
 }
```

4. A hypothetical SARS-CoV-2 Vaccine Trial randomized participants to a vaccine ($T = 1$) or placebo. A safety study followed the participants for one year to determine occurence of adverse events ($Y = 1$) or not. Unfortunately, as is common for studies with a long follow-up period, several participants dropped out of the study early, so that their value for Y was missing. Let $O = 1$ indicate that Y was observed. Suppose participants randomized to placebo and participants experiencing adverse events tended to drop out of the study early. Suppose also that, unknown to the investigators, the vaccine did not cause adverse events. Draw the causal DAG for these three variables, and explain why assessing safety of the vaccine by estimating the association between T and Y for those with $O = 1$ is a biased analysis. This type of bias is referred to as *selection bias*. If you are unsure, you might try a simple simulation in R to obtain the answer.

5. Selection bias also arises from *outcome-based sampling*, such as in a case-control study. With a case-control study design, typically all cases are sampled and only a random fraction of controls. This can be an efficient study design to investigate causes of a rare outcome. For example, prior to the HPV vaccine, case-control studies helped to determine that HPV is a necessary cause of cervical cancer. Let $Y = 1$ indicate cervical cancer, and $T = 1$ indicate infection with HPV. Let $O = 1$ indicate inclusion in the case-control study. Draw

the causal DAG for this case-control study, assuming no unmeasured confounders. The `R` code of `simex5.r` below simulates hypothetical data from this case-control study. Show mathematically and empirically that the risk difference for $O = 1$ is a biased estimate of the risk difference for the entire population represented by `simex5.r`. Therefore, statistical analysis needs to account for the biased sampling. One popular method is to use the odds ratio, which is symmetric in T and Y. Show mathematically that the odds ratio comparing $P(Y = 1|T = 1, O = 1)$ with $P(Y = 1|T = 0, O = 1)$ is identical to that comparing $P(Y = 1|T = 1)$ with $P(Y = 1|T = 0)$. Also show mathematically that when $Y = 1$ is rare in the population, we can interpret the latter odds ratio as a relative risk. How are these results useful?

```
simex5.r <- function ()
 {
   nsim <- 10000
   set.seed(2468)
   probT <- 0.5
   T <- rbinom(n = nsim, size = 1, prob = probT)
   probY <- 0.01 * T
   Y <- rbinom(n = nsim, size = 1, prob = probY)
   probO <- Y + 0.1 * (1 - Y)
   O <- rbinom(n = nsim, size = 1, prob = probO)
   out <- cbind(T, Y, O)
   data.frame(out)
 }
```

6

Adjusting for Confounding: Backdoor Method via Standardization

The methods of this chapter require the assumption that H is a sufficient confounder for estimating the effect of T on Y; that is,

$$\{Y(t)\}_{t \in \mathcal{T}} \amalg T|H, \tag{6.1}$$

where $\mathcal{T}$ is the set of possible values for T, as determined using the Backdoor Theorem explained in Chapter 5. We also require the positivity assumption that

$$1 > P(T = t|H) > 0$$

for all possible values t of T. Finally, we require the consistency assumption. We adopt the terminology for standardization via an outcome model or an exposure model from Rothman et al. (2008).

6.1 Standardization via Outcome Modeling

Standardization estimates a standardized or unconfounded population-average association or causal effect. With it, one can compare groups had the distribution of confounders been identical in both groups to that of the standard population. Standardization via outcome modeling is one way to estimate $E(Y(t))$ assuming (6.1), positivity, and consistency. We can write

$$E(Y(t)) = E_H E(Y(t)|H) =$$
$$E_H E(Y(t)|T = t, H) = E_H E(Y|T = t, H), \tag{6.2}$$

where the first equality follows from the double expectation theorem, the second follows from assumption (6.1), which implies mean independence of $Y(t)$ and T given H, and the third follows from the consistency assumption, which allows us to replace $Y(t)$ with Y when we condition on $T = t$. $E(Y|T = t, H)$ is known as the outcome model. Note that

$$E_H E(Y|T = t, H) \neq E_{H|T=t} E(Y|T = t, H) = E(Y|T = t),$$

DOI: 10.1201/9781003146674-6

where $E_{H|T=t}E(Y|T=t) = E(Y|T=t)$ by the double expectation theorem. Mistakenly equating $E_H E(Y|T=t|H)$ with $E_{H|T=t}E(Y|T=t,H)$ would imply that $E(Y(t)) = E(Y|T=t)$, which would only be true if $Y(t) \amalg T$. However, we have assumed confounding by H; that is, $Y(t) \amalg T|H$.

Equation (6.2) is the basis for the outcome-modeling approach to standardization. If H is continuous or high-dimensional, we cannot express $E(Y|T=t,H)$ nonparametrically, but rather must use a parametric outcome model. With a binary dataset, we can write

$$E_H E(Y|T=t,H) = E(Y|T=t,H=0)P(H=0) + E(Y|T=t,H=1)P(H=1).$$

We can thus estimate $E(Y(t))$ using nonparametric estimates of the components on the right-hand side of the equation. When, for example, we estimate and compare $E(Y(1))$ and $E(Y(0))$, we do so for the distribution of confounders H given by the entire population; that is, for those with $T=1$ and $T=0$, combined.

We consider some examples. For the mortality data of Table 1.1, recalling that $T=1$ indicates the US and $H=1$ indicates the 65+ age group, we have that

$$\begin{aligned}\hat{E}(Y|T=0,H=0) &= 0.002254\\ \hat{E}(Y|T=0,H=1) &= 0.05652\\ \hat{E}(Y|T=1,H=0) &= 0.002679\\ \hat{E}(Y|T=1,H=1) &= 0.04460\end{aligned} \tag{6.3}$$

and

$$\hat{P}(H=0) = \frac{282,305,227 + 1,297,258,493}{282,305,227 + 1,297,258,493 + 48,262,955 + 133,015,479} = 0.897$$
$$\hat{P}(H=1) = 1 - \hat{P}(H=0) = 1 - 0.897 = 0.103$$

Therefore,

$$\begin{aligned}\hat{E}(Y(0)) &= 0.002254 * 0.897 + 0.05652 * 0.103 = 0.0078434\\ \hat{E}(Y(1)) &= 0.002679 * 0.897 + 0.04460 * 0.103 = 0.0069969\end{aligned}$$

Thus, we see that when we combine the age-specific mortality rates of China with the age distribution of the two countries combined, we compute an overall mortality rate of around 7.8 per 1000, whereas when we do the same for the US, we compute about 7.0 per 1000. The standardized risk difference is thus −0.8 per 1000, and the standardized relative risk is 0.897. This is a reversal from the unadjusted comparison of the two countries, which was in the opposite direction, at 8.8 per 1000 versus 7.3 per 1000, with a risk difference of 1.2 per 1000 and a relative risk of 1.206. Due to the large sample sizes per country, the reversal is highly statistically significant.

Next, we use R to compute standardized estimates for the What-If? study analyzed in Chapter 3. We assumed that

$$\{Y(0), Y(1)\} \amalg A|H.$$

We can use the nonparametric linear model

$$E(Y|A, H) = \beta_1 + \beta_2 A + \beta_3 H + \beta_4 A * H$$

to estimate

$$E(Y(0)) = E_H E(Y|A = 0, H) = \beta_1 + \beta_3 E(H)$$

and

$$E(Y(1)) = E_H E(Y|A = 1, H) = \beta_1 + \beta_2 + \beta_3 E(H) + \beta_4 E(H).$$

Using R,

```
bootstand.r <- function ()
  {
    stand.out <- boot(data = whatifdat,
                      statistic = stand.r,
                      R = 1000)
    stand.est <- summary(stand.out)$original
    stand.SE <- summary(stand.out)$bootSE
    stand.lci <- stand.est - 1.96 * stand.SE
    stand.uci <- stand.est + 1.96 * stand.SE
    list(
      stand.est = stand.est,
      stand.SE = stand.SE,
      stand.lci = stand.lci,
      stand.uci = stand.uci
    )
  }
stand.r <- function (data, ids)
  {
    dat <- data[ids,]
    # Find marginal expected value of H
    EH <- mean(dat$H)
    # Fit outcome model
    beta <- lm(Y ~ A * H, data = dat)$coef
    # Compute marginal expected potential outcomes
    EY0 <- beta[1] + beta[3] * EH
    EY1 <- beta[1] + beta[2] + beta[3] * EH + beta[4] * EH
    # Return effect measures
    rd <- EY1 - EY0
    logrr <- log(EY1 / EY0)
    c(EY0, EY1, rd, logrr) .
  }
```

```
> bootstand.r()
$stand.est
[1]  0.375253  0.289004 -0.086248 -0.261158
$stand.SE
[1] 0.057027 0.042122 0.064034 0.192665
$stand.lci
[1]  0.26348  0.20645 -0.21175 -0.63878
$stand.uci
[1] 0.487026 0.371563 0.039258 0.116465
```

The results are presented in Table 6.1, where we have exponentiated the results for the relative risk.

TABLE 6.1
Standardized Estimates for the What-If? Study

Measure	Estimate	95% CI
$\hat{E}(Y(0))$	0.375	(0.263, 0.487)
$\hat{E}(Y(1))$	0.289	(0.206, 0.372)
$\hat{RD}$	−0.086	(−0.212, 0.039)
$\hat{RR}$	0.770	(0.528, 1.12)

We observe that reducing drinking leads to a reduction of unsuppressed viral load in the study population from 37.5% to 28.9%; however, the difference is not statistically significant.

For our final example, we analyze data from the Double What-If? Study. From Chapter 5 and the causal DAG in Figure 5.9, we have that

$$\{VL_1(0), VL_1(1)\} \amalg A|AD_0,$$

but that

$$\{VL_1(0), VL_1(1)\} \not\amalg A|VL_0.$$

From the R code for `doublewhatifsim.r` in Chapter 1 that was used to simulate the data, we can find the true outcome model

$$E(VL_1|A, AD_0).$$

We are given that `VL1prob = VL0prob + .1 - .45*AD1` and `VL0prob= .8-.4*AD0`, so that `VL1prob = .9-.4AD0 - .45*AD1`. This translates to

$$E(VL_1|AD_0, AD_1) = 0.9 - 0.4AD_0 - 0.45AD_1.$$

From the causal DAG of Figure 5.9, we see that VL_1 is independent, and hence mean independent, of A given AD_0 and AD_1. Therefore,

$$E(VL_1|A, AD_0, AD_1) = 0.9 - 0.4AD_0 - 0.45AD_1.$$

By the double expectation theorem, we have that

$$E(VL_1|A, AD_0) = E_{AD_1|AD_0,A}\left(E(VL_1|A, AD_0, AD_1)\right),$$

which equals

$$0.9 - 0.4AD_0 - 0.45E(AD_1|AD_0, A).$$

Again by the causal DAG, we see that AD_1 is mean idependent of AD_0 given A, so that

$$E(VL_1|A, AD_0) = 0.9 - 0.4AD_0 - 0.45E(AD_1|A).$$

Finally, from `doublewhatifsim.r`, we have `AD1prob = 0.1 + 0.8*A`, which translates to

$$E(AD_1|A) = 0.1 + 0.8A.$$

Putting it all together,

$$\begin{aligned} E(VL_1|A, AD_0) &= (0.9 - 0.45 * 0.1) - (0.45 * 0.8) * A - 0.4 * AD_0 \\ &= 0.855 - 0.36A - 0.4AD_0. \end{aligned}$$

We can also find that

$$E(AD_0) = 0.2 + 0.6E(U) = 0.2 + 0.6 * 0.5 = 0.5.$$

Therefore, we can compute the true values of $E(VL_1(0))$ and $E(VL_1(1))$ before we proceed to use the data to estimate them:

$$\begin{aligned} E(VL_1(0)) &= 0.855 - 0.4 * E(AD_0) = 0.855 - 0.4 * 0.5 = 0.655 \\ E(VL_1(1)) &= 0.855 - 0.36 - 0.4 * E(AD_0) = 0.855 - 0.36 - 0.4 * 0.5 = 0.295 \end{aligned}$$

For estimation, we can reuse the R code used for the What-If? Study, labeling $Y = VL_1$ and $H = AD_0$. The results are presented in Table 6.2.

TABLE 6.2
Standardized Estimates for the Double What-If? Study with $H = AD_0$

Measure	Estimate	95% CI
$\hat{E}(VL_1(0))$	0.669	(0.636, 0.702)
$\hat{E}(VL_1(1))$	0.335	(0.272, 0.397)
$\hat{RD}$	−0.334	(−0.403, −0.265)
$\hat{RR}$	0.500	(0.413, 0.606)

We observe that the confidence interval for $E(VL_1(0))$, which is (0.636, 0.702), includes the true value 0.655; the confidence interval for $E(VL_1(1))$, which is (0.272,0.397), includes the true value 0.295. As expected, the method appears to work. However, without knowledge of the foregoing proof that the method does indeed work, we might wonder if we were just lucky with this particular sample, which we obtained with `set.seed(444)`. More empirical evidence that the method works can be obtained by repeatedly running `doublewhatifsim.r` without the `set.seed(444)` command in order to generate 1000 confidence intervals; we can then ascertain whether the percentage of confidence intervals covering the true values is 95%. We implement this in R as follows:

```
empiricalsim.r <- function()
  {
    # Initialize coverage indicators
    outY0 <- rep(0, 1000)
    outY1 <- rep(0, 1000)
    for (i in 1:1000)
    {
      # Generate a random dataset
      dat <- doublewhatifsim.r()
      dubs <- dat
      # Relabel the variables for estimation with bootstand.r
      dubs$H <- dubs$AD0
      dubs$Y <- dubs$VL1
      # Find the estimates
      est <- bootstand.r(datain = dubs)
      # Check the coverage
      if ((est$stand.lci[1] < .655) & (est$stand.uci[1] > .655))
        outY0[i] <- 1
      if ((est$stand.lci[2] < .295) & (est$stand.uci[2] > .295))
        outY1[i] <- 1
    }
    list(coverage0 = mean(outY0), coverage1 = mean(outY1))
  }
empircalsim.r()
$coverage0
[1] 0.944
$coverage1
[1] 0.951
```

The percentage of confidence intervals for $E(VL_1(0))$ covering 0.655 is 94.4%, while the percentage of confidence intervals for $E(VL_1(1))$ covering 0.295 is 95.1%. These numbers are quite close to 95%. They are not exactly equal to 95% due to sampling variability from only 1000 sampled datasets from `doublewhatifsim.r` and only 1000 bootstrap samples. This simulation study gives us more evidence that the method is valid. However, one simulation study is not a general proof. Fortunately, we have already proved that the method works.

For comparison, we repeat the standardization with $H = VL_0$. The results are presented in Table 6.3.

TABLE 6.3
Standardized Estimates for the Double What-If? Study with $H = VL_0$

Measure	Estimate	95% CI
$\hat{E}(VL_1(0))$	0.696	(0.662, 0.729)
$\hat{E}(VL_1(1))$	0.245	(0.192, 0.299)
$\hat{RD}$	−0.450	(−0.512, −0.388)
$\hat{RR}$	0.353	(0.283, 0.441)

As VL_0 is not a sufficient confounder, we do not expect these results to be correct. Indeed, the confidence interval for $E(VL_1(0))$, which is (0.662, 0.729), does not include the true $E(VL_1(0))$, which is 0.655. However, the confidence interval for $E(VL_1(1))$, (0.272, 0.397), does include the true $E(VL_1(1))$, which is 0.295. Therefore, the results are biased, but not terribly so. Furthermore, due to sampling variability, it is possible that the method is valid but that our confidence interval belongs to the 5% of the 95% confidence intervals that do not contain the true value. We conducted a simulation study using `R` code analogous to `empiricalsim.r` and found that the percentage of confidence intervals for $E(VL_1(0))$ covering 0.655 is 63.3%, while the percentage of confidence intervals for $E(VL_1(1))$ covering 0.295 is 13.7%. Because these percentages are so far away from 95%, the simulation study indicates substantial bias.

It is important to keep in mind that this is only one simulation study, and that other simulation studies with different data generating mechanisms might reveal more or less bias. However, just one simulation study demonstrating bias is enough to prove that standardization with an insufficient confounder, such as VL_0, is not generally valid.

6.1.1 Average Effect of Treatment on the Treated

In Chapter 3, we introduced a special kind of conditional causal effect, called the *average effect of treatment on the treated* (ATT), i.e. $E(Y(1)|A = 1)$ versus $E(Y(0)|A = 1)$. We can estimate $E(Y(1)|A = 1)$ easily, because it equals $E(Y|A = 1)$ by consistency. We can use an outcome-modeling approach to standardization to estimate $E(Y(0)|A = 1)$, assuming (6.1), with A in place of T, and consistency:

$$E(Y(0)|A = 1) = E_{H|A=1}E(Y(0)|A = 1, H) =$$
$$E_{H|A=1}E(Y(0)|A = 0, H) = E_{H|A=1}E(Y|A = 0, H), \tag{6.4}$$

where the first equality follows from the double expectation theorem, the second from (6.1), and the third from consistency. For binary H, we can estimate the outcome model non-parametrically, as before, but for continuous or high-dimensional H, we can enlist a parametric model for $E(Y|A = 0, H)$. We can use a similar argument to estimate $E(Y(1)|A = 0)$; this is left as an exercise.

For the mortality data, letting $A = T$, the ATT compares mortality in the US versus China using the age distribution in the US; i.e. the US serves as the standard population. Recalling from Chapter 1 that $P(H = 1|A = 1) = 0.146$, so that $P(H = 0|A = 1) = 0.854$, and using the mortality rates at (6.3), we compute $E(Y(0)|A = 1)$ from (6.4) as

$$\begin{gathered}E(Y|A=0,H=0)P(H=0|A=1)+\\E(Y|A=0,H=1)P(H=1|A=1)=\\0.002254*0.854+0.05652*0.146=0.0102.\end{gathered}$$

Thus, had the age distribution been the same as in the US, the mortality rate in China would have been 10.2 per 1000 instead of 7.3 per 1000. We can compute $E(Y(1)|A = 1)$ directly from the US data as $E(Y|A = 1) = 0.0088$, or 8.8 per 1000. The ATT is thus 8.8 per 1000 versus 10.2 per 1000.

For the What-If? study, we use R to estimate the ATT, with a slight modification to our previous code.

```
bootstandatt.r <- function ()
{
  stand.out <- boot(data = whatifdat,
                    statistic = standatt.r,
                    R = 1000)
  stand.est <- summary(stand.out)$original
  stand.SE <- summary(stand.out)$bootSE
  stand.lci <- stand.est - 1.96 * stand.SE
  stand.uci <- stand.est + 1.96 * stand.SE
  list(
    stand.est = stand.est,
    stand.SE = stand.SE,
    stand.lci = stand.lci,
    stand.uci = stand.uci
  )
}
standatt.r <- function (data, ids)
{
  dat <- data[ids, ]
  # Compute the expected value of H given A=1
  EHA <- mean(dat$H[dat$A == 1])
  # Fit the outcome model
  beta <- lm(Y ~ A * H, data = dat)$coef
  # Compute the expected potential outcomes given A=1
```

```
    EY0A <- beta[1] + beta[3] * EHA
    EY1A <- beta[1] + beta[2] + beta[3] * EHA + beta[4] * EHA
    rd <- EY1A - EY0A
    logrr <- log(EY1A / EY0A)
    c(EY0A, EY1A, rd, logrr)
  }
> bootstandatt.r()
$stand.est
[1]  0.361111  0.276190 -0.084921 -0.268095
$stand.SE
[1] 0.058430 0.042816 0.062080 0.196675
$stand.lci
[1]  0.24659  0.19227 -0.20660 -0.65358
$stand.uci
[1] 0.475633 0.360109 0.036756 0.117387
```

TABLE 6.4
ATT for the What-If? Study

Measure	Estimate	95% CI
$\hat{E}(Y(0)\|A=1)$	0.361	(0.247, 0.476)
$\hat{E}(Y(1)\|A=1)$	0.276	(0.192, 0.360)
$\hat{RD}$	−0.085	(−0.207, 0.037)
$\hat{RR}$	0.765	(0.520, 1.12)

In this example, we find that the results reported in Table 6.4 are very similar to the overall average effect of treatment, $E(Y(1)) - E(Y(0))$, computed previously, although the results for both are variable due to the relatively small sample size. This suggests that any effect modifiers are balanced across the two treatment groups.

For comparison with difference-in-differences estimation presented in Chapter 7, we estimate the ATT for the Double What-If? Study, first with $H = AD_0$ and second with $H = VL_0$. The analysis with $H = AD_0$ is correct, whereas the one with $H = VL_0$ is not. Computing the true values of $E(VL_1(1)|A = 1)$, $E(VL_1(0)|A = 1)$, and the ATT using the true data generating mechanisms in `doublewhatifsim.r` is difficult but not impossible. Previously, we found that

$$E(VL_1|A, AD_0) = 0.855 - 0.36A - 0.4AD_0.$$

From (6.4), we can compute

$$E(VL_1(0)|A=1) = E_{AD_0|A=1}E(VL_1|A=0, AD_0) = \\ E_{AD_0|A=1}(0.855 - 0.4AD_0) = 0.855 - 0.4 * E(AD_0|A=1). \qquad (6.5)$$

We can compute

$$E(AD_0|A=1) = P(AD_0=1|A=1) = \frac{P(A=1|AD_0=1)P(AD_0=1)}{P(A=1)}. \tag{6.6}$$

Previously, we found that $E(AD_0) = 0.5$; this also equals $P(AD_0 = 1)$. From `doublewhatifsim.r`, we have that `Aprob = 0.05+T*U*0.8`, which translates to

$$E(A|T,U) = 0.05 + T * U * 0.8.$$

Therefore, $P(A = 1)$ equals

$$E(A) = E_{T,U}E(A|T,U) = 0.05 + E(T * U) * 0.8,$$

and since by `doublewhatifsim.r` T and U are independent each with expected values equal to 0.5, $E(T * U) = E(T) * E(U) = 0.5 * 0.5 = 0.25$. Thus

$$P(A = 1) = 0.05 + 0.25 * 0.8 = 0.25.$$

Next, we need to compute $P(A = 1|AD_0 = 1)$. We have that

$$P(A=1|AD_0=1) = P(A=1|U=0,AD_0=1)P(U=0|AD_0=1) + P(A=1|U=1,AD_0=1)P(U=1|AD_0=1). \tag{6.7}$$

We also have

$$P(U|AD_0 = 1) = P(AD_0 = 1|U)P(U)/P(AD_0 = 1).$$

From the causal DAG of Figure 5.9, $A \amalg AD_0|U$, so that

$$P(A = 1|U, AD_0 = 1) = P(A = 1|U).$$

From `doublewhatifsim.r`, we have

$$P(AD_0 = 1|U) = 0.2 + 0.6U.$$

We have

$$P(AD_0 = 1) = P(AD_0 = 1|U = 0)P(U = 0) + P(AD_0 = 1|U = 1)P(U = 1).$$

And from `doublewhatifsim.r`, we have $P(U = 1) = 0.5$. Therefore,

$$P(AD_0 = 1) = 0.2 * 0.5 + 0.8 * 0.5 = 0.5,$$

and

$$P(U = 0|AD_0 = 1) = 0.2 * 0.5/0.5 = 0.2$$
$$P(U = 1|AD_0 = 1) = 0.8 * 0.5/0.5 = 0.8.$$

We have that

$$P(A=1|U) = P(A=1|U,T=0)P(T=0|U)+P(A=1|U,T=1)P(T=1|U),$$

which gives us

$$P(A=1|U=0) = 0.05 * 0.5 + 0.05 * 0.5 = 0.05$$
$$P(A=1|U=1) = 0.05 * 0.5 + 0.85 * 0.5 = 0.45$$

Putting everything into equation (6.7), we have

$$P(A=1|AD_0=1) = 0.05 * 0.2 + 0.45 * 0.8 = 0.37,$$

so that, by equation (6.6),

$$E(AD_0|A=1) = 0.37 * 0.5/0.25 = 0.74.$$

Finally, by equation (6.5),

$$E(VL_1(0)|A=1) = 0.855 - 0.4 * 0.74 = 0.559.$$

We can compute $E(VL_1(1)|A=1)$ as follows:

$$\begin{aligned} E(VL_1|A=1) = E_{AD_0|A=1}E(VL_1|A=1, AD_0) = \\ 0.855 - 0.36 - 0.4E(AD_0|A=1) = \\ 0.855 - 0.36 - 0.4 * 0.74 = 0.199. \end{aligned}$$

Thus, we have $E(VL_1(0)|A=1) = 0.559$, $E(VL_1(1)|A=1) = 0.199$, the true RD equal to $E(VL_1(1)|A=1) - E(VL_1(0)|A=1) = -0.36$, and the true RR equal to $E(VL_1(1)|A=1)/E(VL_1(0)|A=1) = 0.356$.

We can reuse the R code `standatt.r` and `bootstandatt.r`, labeling $Y = VL_1$ and $H = AD_0$. The results are presented in Table 6.5.

TABLE 6.5
Standardized ATT Estimates for the Double What-If? Study with $H = AD_0$

Measure	Estimate	95% CI
$\hat{E}(VL_1(0)\|A=1)$	0.574	(0.526, 0.622)
$\hat{E}(VL_1(1)\|A=1)$	0.231	(0.179, 0.283)
$\hat{RD}$	−0.344	(−0.404, −0.283)
$\hat{RR}$	0.402	(0.322, 0.501)

We observe that the confidence interval for $E(VL_1(0)|A=1)$, which is (0.526, 0.622), includes the true value 0.559, and the one for $E(VL_1(1)|A=1)$, which is (0.179, 0.283), includes the true value 0.199. The true RD, −0.36, is in its respective confidence interval (−0.404, −0.283), and the true RR, 0.356, also falls in its respective confidence interval (0.322, 0.501). As expected, the method appears to work. We conducted a simulation study with R code analogous to `empiricalsim.r` to find that the confidence intervals cover $E(Y(0)|A=1)$ 93.7% of the time, $E(Y(0)|A=1)$ 94.7% of the time, the true RD 95.3% of the time, and the true RR 96% of the time. This supports the validity of the method.

The results for $H = VL_0$ are presented in Table 6.6.

TABLE 6.6
Standardized ATT Estimates for the Double What-If? Study with $H = VL_0$

Measure	Estimate	95% CI
$\hat{E}(VL_1(0)\|A=1)$	0.682	(0.647, 0.718)
$\hat{E}(VL_1(1)\|A=1)$	0.231	(0.177, 0.284)
$\hat{RD}$	−0.452	(−0.514, −0.389)
$\hat{RR}$	0.338	(0.267, 0.428)

As VL_0 is not a sufficient confounder, we do not expect these results to be correct. Indeed, the only confidence intervals that include the true values are for $E(VL_1(1)|A=1)$ and RR. For the other two measures, the results are biased, but not terribly so. We should not expect the confidence interval for $E(VL_1(1)|A=1)$ and the RR to cover the true value in general. Once again, we conducted a simulation study and found that the confidence intervals cover $E(VL_1(0)|A=1)$ 0% of the time, $E(VL_1(1)|A=1)$ 96.0% of the time, the true RD 7.7% of the time, and the true RR 75.5% of the time. Therefore, for the data generating model of `doublewhatifsim.r`, use of VL_0 rather than AD_0 as a sufficient confounder is invalid for three of the four measures.

6.1.2 Standardization with a Parametric Outcome Model

When H necessitates a parametric outcome model, validity of the standardization depends upon correct model specification. In practice, our model will never be exactly right, but hopefully it is close enough that our results are not too far off. We can make slight modifications to `stand.r` and `bootstand.r` to compute the estimated potential outcomes using equation (6.2). We present two examples. First, we analyze the What-If? data in `whatif2dat`, which includes some continuous covariates.

```
> head(whatif2dat)
  vl0 vlcont0 artad0 vl4 vlcont4 artad4 audit0 T A lvlcont0 lvlcont4
1   0      20      1   1     420      1      1 0 1   2.9957   6.0403
2   0      20      0   0      20      1      1 1 1   2.9957   2.9957
```

```
3   1  61420   1  0      20   1   1 1 0  11.0255   2.9957
4   1    600   0  0      20   1   1 1 1   6.3969   2.9957
5   1  75510   0  1  184420   0   1 0 1  11.2320  12.1250
6   0     20   1  0      30   0   1 1 0   2.9957   3.4012
```

The variable `lvlcont0` represents log of viral load (copies/ml) at baseline, a continuous variable that we will use as H in the analysis of the effect of reduced drinking `A` on unsuppressed viral load at four months `vl4`. The following `R` code, together with the bootstrap, computes the estimates shown in Table 6.7:

```
standout.r <- function(data = whatif2dat,
                       ids = c(1:nrow(whatif2dat)))
  {
    dat <- data[ids,]
    # Fit the parametric outcome model
    lmod <- glm(vl4 ~ A + lvlcont0, family = binomial, data = dat)
    dat0 <- dat1 <- dat
    # Create a dataset with everyone untreated
    dat0$A <- 0
    # Create a dataset with everyone untreated
    dat1$A <- 1
    # Compute the expected potential outcome for each
    # participant if untreated
    EYhat0 <- predict(lmod, newdata = dat0, type = "response")
    # Compute the expected potential outcomes for each
    participant if treated
    EYhat1 <- predict(lmod, newdata = dat1, type = "response")
    # Estimate the average potential outcomes
    EY0 <- mean(EYhat0)
    EY1 <- mean(EYhat1)
    # Estimate the effect measures
    rd <- EY1 - EY0
    logrr <- log(EY1 / EY0)
    c(EY0, EY1, rd, logrr)
  }
> bootstand.r()
$stand.est
[1]  0.360421  0.299677 -0.060743 -0.184566
$stand.SE
[1] 0.056824 0.042900 0.064995 0.198183
$stand.lci
[1]  0.24905  0.21559 -0.18813 -0.57300
$stand.uci
[1] 0.471796 0.383762 0.066647 0.203873
```

We observe that the results of Table 6.1 based on the nonparametric outcome model with H equal to unsuppressed viral load at baseline are quite similar

TABLE 6.7
Outcome-model Standardization for the What-If? Study with H = `lvlcont0`

Measure	Estimate	95% CI
$\hat{E}(Y(0))$	0.360	(0.249, 0.472)
$\hat{E}(Y(1))$	0.300	(0.216, 0.384)
$\hat{RD}$	−0.061	(−0.188, 0.067)
$\hat{RR}$	0.831	(0.564, 1.23)

to those of Table 6.7 based on the parametric outcome model with H equal to the log of viral load at baseline.

Second, we analyze the General Social Survey data `gssrcc`, focusing, as in Chapter 3 (see Tables 3.2 and 3.3), on the effect of completing more than high school on reporting a vote for Trump in 2016. Although H contains only binary variables, there are too many of them to use a nonparametric outcome model, which would include all two-way and higher-order interactions in addition to the main effects. It is implausible that we have included a sufficient set of confounders in the model; however, the example nicely illustrates some of the methods in this chapter. We use the following modification to `standout.r`, together with the bootstrap:

```
standout.r <- function(data, ids)
{
  dat <- data[ids, ]
  # Fit the parametric outcome model
  lmod <-
    glm(trump ~ gthsedu + magthsedu + white + female + gt65,
        family = binomial,
        data = dat)
  dat0 <- dat1 <- dat
  dat0$gthsedu <- 0
  dat1$gthsedu <- 1
  # Compute expected potential outcomes for each participant
  EYhat0 <- predict(lmod, newdata = dat0, type = "response")
  EYhat1 <- predict(lmod, newdata = dat1, type = "response")
  # Estimate expected potential outcomes
  EY0 <- mean(EYhat0)
  EY1 <- mean(EYhat1)
  # Estimate effect meausres
  rd <- EY1 - EY0
  rr <- EY1 / EY0
  or <- (EY1 / (1 - EY1)) / (EY0 / (1 - EY0))
  c(EY0, EY1, rd, rr, or)
}
```

The use of the `predict` function together with the output of the `glm` function simplifies the programming for models with more covariates. The standardized estimates are in Table 6.8.

TABLE 6.8
Outcome-model Standardization of the Effect of More than High School Education on Voting for Trump

Measure	Estimate	95% CI
$\hat{E}(Y(0))$	0.233	(0.210, 0.256)
$\hat{E}(Y(1))$	0.271	(0.242, 0.300)
$\hat{RD}$	0.038	(0.001, 0.075)
$\hat{RR}$	1.164	(1.006, 1.347)

We observe that the estimates of $E(Y(0))$ and $E(Y(1))$ are almost identical to those of $E(Y|T=0)$, estimated at 0.233 (0.210, 0.256), and $E(Y|T=1)$, estimated at 0.271 (0.241, 0.301), previously reported in Chapter 3. Furthermore, the estimates of RD and RR are similar to the estimates reported in Table 3.2. We will revisit this example in the next section.

6.2 Standardization via Exposure Modeling

Standardization via exposure modeling gives us a second way to estimate $E(Y(t))$ assuming (6.1), positivity, and consistency. We present the method for binary T. To estimate $E(Y(t))$ for non-binary T, one can first recode the data so that $T=1$ when it previously equaled t, and $T=0$ when it previously equaled any value other than t. Then one can use the method for estimating $E(Y(1))$ with the recoded data.

The exposure model is $E(T|H)=P(T=1|H)$, so named because sometimes T indicates a potentially harmful exposure, rather than a treatment. The exposure model is also known as the *propensity score* (Rosenbaum and Rubin (1983)), denoted by $e(H)$, as it is a function of H. It is called the propensity score because it measures the propensity for treatment given observed levels of the confounders H. The most common parametric model for it is the logistic model

$$E(T|H)=\text{expit}(\alpha_0+H_1\alpha_1+\cdots+H_q\alpha_q), \tag{6.8}$$

where q is the number of components of H.

We will prove the following relations, which we can use to estimate $E(Y(0))$ and $E(Y(1))$.

$$E(Y(1)) = E\left(\frac{TY}{e(H)}\right) \tag{6.9}$$

and

$$E(Y(0)) = E\left(\frac{(1-T)Y}{1-e(H)}\right). \tag{6.10}$$

The proof of (6.9) for discrete Y and H follows. The proof is analogous for (6.10), noting that $P(T=0|H) = 1 - P(T=1|H) = 1 - e(H)$.

$$\begin{aligned}
E\left(\tfrac{TY}{e(H)}\right) &= \Sigma_{y,t,h}\tfrac{ty}{e(h)}P(Y=y,T=t,H=h)\\
&= \Sigma_{y,t,h}\tfrac{ty}{e(h)}P(Y=y|T=t,H=h)P(T=t|H=h)P(H=h)\\
&= \Sigma_{y,t,h}\tfrac{ty}{e(h)}P(Y=y|T=1,H=h)e(h)P(H=h)\\
&= \Sigma_{y,h}\tfrac{y}{e(h)}P(Y=y|T=1,H=h)e(h)P(H=h)\\
&= \Sigma_{y,h}yP(Y=y|T=1,H=h)P(H=h)\\
&= E_H(E(Y|T=1,H)) = E(Y(1))
\end{aligned}$$

The first equality follows from the definition of expectation, the second from the multiplication rule, the third by the definition of $e(H)$, the fourth because when $t = 0$ the summand is zero, the fifth by cancellation of $e(h)$ from numerator and denominator, the sixth and seventh by (6.2). When Y and/or H are continuous, the proof is analogous using integrals and probability density functions in place of sums and probability mass functions.

For better understanding, we will estimate $E(Y(1))$ and $E(Y(0))$ using the mortality data of Table 1.1 and empirical versions of (6.9) and (6.10), and compare with the outcome-modeling standardization results.

```
mk.mortdat <- function() {
    mortdat <- NULL
    # Create columns of the dataset
    mortdat$H <- c(0, 0, 0, 0, 1, 1, 1, 1)
    mortdat$T <- c(0, 0, 1, 1, 0, 0, 1, 1)
    mortdat$Y <- c(0, 1, 0, 1, 0, 1, 0, 1)
    mortdat$n <- c((1297258493 - 2923480),
                   2923480,
                   (282305227 - 756340),
                   756340,
                   (133015479 - 7517520),
                   7517520,
                   (48262955 - 2152660),
                   2152660
    )
    # Compute the proportions who died
    mortdat$p <- mortdat$n / sum(mortdat$n)
    # Compute e(H=0)
    eH0 <- sum(mortdat$n[3:4]) / sum(mortdat$n[1:4])
```

```
    # Compute e(H=1)
    eH1 <- sum(mortdat$n[7:8]) / sum(mortdat$n[5:8])
    # Compute e(H) for all participants
    mortdat$eH <- eH0 * (1 - mortdat$H) + eH1 * mortdat$H
    # Compute the summands of the estimating equation
    mortdat$s1 <- mortdat$T * mortdat$Y / mortdat$eH
    mortdat$s0 <- (1 - mortdat$T) * mortdat$Y / (1 - mortdat$eH)
    # Estimate the expected values of the potential outcomes
    EY1 <- sum(mortdat$s1 * mortdat$p)
    EY0 <- sum(mortdat$s0 * mortdat$p)
    mortdat <- data.frame(mortdat)
    list(EY1 = EY1,
         EY0 = EY0,
         mortdat = mortdat)
  }
> mortdat.out<-mk.mortdat()
> mortdat.out
$EY1
[1] 0.0069952
$EY0
[1] 0.0078399
$mortdat
  H T Y          n          p      eH     s1     s0
1 0 0 0 1294335013 0.73506589 0.17872 0.0000 0.0000
2 0 0 1    2923480 0.00166027 0.17872 0.0000 1.2176
3 0 1 0  281548887 0.15989445 0.17872 0.0000 0.0000
4 0 1 1     756340 0.00042953 0.17872 5.5952 0.0000
5 1 0 0  125497959 0.07127156 0.26624 0.0000 0.0000
6 1 0 1    7517520 0.00426928 0.26624 0.0000 1.3628
7 1 1 0   46110295 0.02618650 0.26624 0.0000 0.0000
8 1 1 1    2152660 0.00122252 0.26624 3.7561 0.0000
```

In the function `mk.mortdat`, which makes the `mortdat` dataset and calculates the estimated potential outcomes, `eH` is $e(H)$, `s1` is $TY/e(H)$, and `s0` is $(1-T)Y/(1-e(H))$. To compute the estimated potential outcomes, we need to sum the `s1` or `s0` summands weighted by the probability of the row, `p`. We see that $\hat{E}(Y(1)) = 0.0069952$ and $\hat{E}(Y(0)) = 0.0078399$, which would be identical to the estimates obtained earlier using the outcome-modeling approach, except for propagation of round-off error.

6.2.1 Average Effect of Treatment on the Treated

We can also use the exposure-modeling approach to estimate the ATT. We introduce $e_0 = P(T = 1)$, to go along with $e(H) = P(T = 1|H)$. We showed in equation (6.4), letting $T = A$, that

$$E(Y(0)|T = 1) = E_{H|T=1}E(Y|T = 0, H). \tag{6.11}$$

We will prove that $E(Y(0)|T=1)$ is also a function of the exposure model. Specifically,

$$E(Y(0)|T=1) = E\left(\frac{Y(1-T)e(H)}{(1-e(H))e_0}\right). \tag{6.12}$$

For the proof, we have that

$$\begin{aligned}
E_{H|T=1}E(Y|T=0,H) &= E_{H|T=1}\Sigma_y yP(Y=y|T=0,H) \\
&= \Sigma_h\Sigma_y yP(Y=y|T=0,H=h)P(H=h|T=1) \\
&= \Sigma_h\Sigma_y yP(Y=y|T=0,H=h)P(T=1|H=h)P(H=h)/P(T=1) \\
&= \Sigma_{h,y}\tfrac{yP(T=1|H=h)}{P(T=1)}\tfrac{P(Y=y|T=0,H=h)P(T=0|H=h)P(H=h)}{P(T=0|H=h)} \\
&= \Sigma_{h,y}\tfrac{yP(T=1|H=h)}{P(T=0|H=h)P(T=1)}P(Y=y,T=0,H=h) \\
&= \Sigma_{h,y,t}\tfrac{y(1-t)e(h)}{(1-e(h))e_0}P(Y=y,T=t,H=h) \\
&= E\left(\tfrac{Y(1-T)e(H)}{(1-e(H))e_0}\right)
\end{aligned}$$

The first and second equalities follows from the definition of conditional expectation and double expectation, the third from the multiplication rule, the fourth from $P(T=0|H=h)/P(T=0|H=h)=1$ and rearranging terms, the fifth from the multiplication rule and rearranging terms, the sixth from the definition of $e(H)$ and e_0 and from the equality $\Sigma_t(1-t)P(Y=y,T=t,H=h)=P(Y=y,T=0,H=h)$, and the seventh from the definition of expectation. We can use a similar argument to estimate $E(Y(1)|T=0)$; this is left as an exercise.

We compute $E(Y(0)|T=1)$ as follows

```
attsem.r <- function(mortdat = mortdat.out$mortdat)
{
  # Estimate P(T=1)
  e0 <- sum(mortdat$T * mortdat$p)
  # Compute the summands of the estimating equation
  s <- mortdat$Y * (1 - mortdat$T) * mortdat$eH /
      (e0 * (1 - mortdat$eH))
  # Estimate E(YO|T=1)
  EYOT1 <- sum(s * mortdat$p)
  EYOT1
}
> attsem.r()
[1] 0.010176
```

We find that $\hat{E}(Y(0)|T=1)=0.0102$, or 10.2 per 1000, identical to the estimate computed using the outcome-modeling approach. $E(Y(1)|T=1)$ can be estimated via $E(Y|T=1)$, as before, at 8.8 per 1000.

6.2.2 Standardization with a Parametric Exposure Model

For parametric exposure models such as (6.8), equation (6.9) allows us to estimate $E(Y(1))$ by first estimating α with $\hat{\alpha}$ using the estimating equation given in Chapter 2 for logistic models, second computing $\hat{e}(H)$ with $\hat{\alpha}$, and third computing

$$\hat{E}(Y(1)) = \frac{1}{n}\Sigma_i \frac{T_i Y_i}{\hat{e}(H_i)}. \tag{6.13}$$

Similarly, we can estimate $E(Y(0))$ as

$$\hat{E}(Y(0)) = \frac{1}{n}\Sigma_i \frac{(1-T_i)Y_i}{1-\hat{e}(H_i)}. \tag{6.14}$$

Pretending that $Y(1) = 1$ for everyone, the proof of (6.9) implies that

$$1 = E(1) = E\left(\frac{T}{e(H)}\right),$$

so that another good estimator of $E(Y(1))$ is given by

$$\hat{E}(Y(1)) = \frac{\Sigma_i \frac{T_i Y_i}{\hat{e}(H_i)}}{\Sigma_i \frac{T_i}{\hat{e}(H_i)}},$$

because the expectation of the denominator is n. This takes the form of a random *weighted average*,

$$\begin{aligned} \hat{E}(Y(1)) &= \Sigma_i W_i Y_i, \quad \text{where} \\ W_i &= \frac{\frac{T_i}{\hat{e}(H_i)}}{\Sigma_i \frac{T_i}{\hat{e}(H_i)}}, \end{aligned} \tag{6.15}$$

where $W_i \in [0, 1]$ and $\Sigma_i W_i = 1$. Note that $(1/n)\Sigma_i Y_i$ is a weighted average with $W_i = 1/n$. Similarly,

$$\frac{\Sigma_i a_i Y_i}{\Sigma_i a_i}$$

is a weighted average for any nonnegative a_i. We also have that

$$\begin{aligned} \hat{E}(Y(0)) &= \Sigma_i W_i Y_i, \quad \text{where} \\ W_i &= \frac{\frac{1-T_i}{1-\hat{e}(H_i)}}{\Sigma_i \frac{1-T_i}{1-\hat{e}(H_i)}}. \end{aligned} \tag{6.16}$$

The estimators of (6.15) and (6.16) are useful for estimation using the `weighted.mean` function or the `glm` or `geeglm` functions with the `weights` options. Note that the denominators of the weights are constant for a given sample, and our use of the three `R` functions is invariant to multiplication of the weights by such a constant. Therefore, in our data examples, we use only the numerators for the weights.

First, we apply exposure-model standardization to the What-If? Study with H = `lvlcont0`, for comparison with the outcome-modeling results reported in Table 6.7. We use the `standexp.r` function shown below, noting that the use of the `weights` option in the `glm` function produces the correct estimator of `beta`, but that the standard errors are incorrect for our usage. Therefore, we turn to the bootstrap, which is also convenient for computing confidence intervals for functions of the parameters, including the relative risk. Some readers may additionally find the `twang` package in `R` helpful for exposure-model standardization and related objectives.

```
standexp.r <- function(data, ids)
  {
    dat <- data[ids, ]
    # Estimate the parametric exposure model
    e <- fitted(glm(A ~ lvlcont0, family = binomial, data = dat))
    # Compute the weights
    dat$W <- (1 / e) * dat$A + (1 / (1 - e)) * (1 - dat$A)
    # Fit the weighted linear model
    beta <- glm(vl4 ~ A, data = dat, weights = W)$coef
    # Estimate the expected potential outcomes
    EY0 <- beta[1]
    EY1 <- beta[1] + beta[2]
    # Estimate the effect measures
    rd <- EY1 - EY0
    rr <- log(EY1 / EY0)
    c(EY0, EY1, rd, rr)
  }
> bootstand.r()
$stand.est
[1]  0.3600  0.3003 -0.0598 -0.1816
$stand.SE
[1] 0.0567 0.0408 0.0655 0.1989
$stand.lci
[1]  0.249  0.220 -0.188 -0.571
$stand.uci
[1] 0.4712 0.3803 0.0687 0.2082
```

The results are reported in Table 6.9, which is almost unchanged from Table 6.7.

TABLE 6.9
Exposure-model Standardization for the What-If? Study with H = `lvlcont0`

Measure	Estimate	95% CI
$\hat{E}(Y(0))$	0.360	(0.249, 0.471)
$\hat{E}(Y(1))$	0.300	(0.220, 0.380)
$\hat{RD}$	−0.060	(−0.188, 0.069)
$\hat{RR}$	0.834	(0.565, 1.23)

If interest is focused primarily on the risk difference, one could alternatively program its estimator and standard error using `exp.r`:

```
  exp.r <- function(dat = whatif2dat)
    library(geepack)
  # Fit the parametric exposure model
  e <- fitted(glm(A ~ lvlcont0, family = binomial, data = dat))
  # Compute the weights
  dat$W <- (1 / e) * dat$A + (1 / (1 - e)) * (1 - dat$A)
  # Provide participant ids to geeglm
  dat$ids <- c(1:nrow(dat))
  # Fit the weighted linear model using geeglm
  summary(geeglm(
    vl4 ~ A,
    data = dat,
    id = ids,
    weights = W
  ))
}
Call:
geeglm(formula = vl4 ~ A, data = dat, weights = W, id = ids)

 Coefficients:
            Estimate Std.err  Wald Pr(>|W|)
(Intercept)   0.3600  0.0613 34.47  4.3e-09
A            -0.0598  0.0767  0.61     0.44
```

We see that the coefficient of `A`, -0.0598, is identical to our estimate of the RD in Table 6.9. It has been shown that for large samples, the standard error, 0.0767, of the RD estimated with `exp.r` is necessarily larger than our bootstrap estimate of the standard error estimated with `bootstand.r` and `standexp.r`. As the latter standard error is 0.0655, the rule holds for this example. Because the standard error estimated with `exp.r` is too large, if one were to find statistical signficance, typically $P < 0.05$, then one would not need to do any further analyses, because the bootstrap P-value would be even smaller. However, if $P > 0.05$, as it does in this example with $P = 0.44$, one would typically wish to check statistical signficance with the bootstrap. Our bootstrap confidence interval includes 0, which means that for this example, both methods produce statistically insignificant results.

Next we turn to the General Social Survey data `gssrcc`, comparing standardized estimates with an exposure model to our outcome-modeling estimates in Table 6.8. We modify `standexp.r` as follows, and we again use `bootstand.r`.

```
standexp.r <- function(data, ids)
{
  dat <- data[ids, ]
  # Fit the parametric exposure model
  e <-
```

```
    fitted(glm(
      gthsedu ~ magthsedu + white + female + gt65,
      family = binomial,
      data = dat
    ))
  # Compute the weights
  dat$W <- (1 / e) * dat$gthsedu + (1 / (1 - e)) * (1 - dat$gthsedu)
  # Fit the weighted linear model
  beta <- glm(trump ~ gthsedu, data = dat, weights = W)$coef
  # Estimate the expected potential outcomes
  EY0 <- beta[1]
  EY1 <- beta[1] + beta[2]
  # Estimate the effect measures
  rd <- EY1 - EY0
  rr <- log(EY1 / EY0)
  c(EY0, EY1, rd, rr)
}
```

The results are shown in Table 6.10.

TABLE 6.10
Exposure-model Standardization of the Effect of More than High School Education on Voting for Trump

Measure	Estimate	95% CI
$\hat{E}(Y(0))$	0.231	(0.208, 0.254)
$\hat{E}(Y(1))$	0.272	(0.242, 0.301)
$\hat{RD}$	0.041	(0.003, 0.078)
$\hat{RR}$	1.176	(1.015, 1.361)

We observe that the outcome-model standardization estimates in Table 6.8 are almost identical to the exposure-model standardization estimates in Table 6.10, which we already noted were similar to the unadjusted estimates reported in and above Table 3.2. The similarities suggest that either the arrow from H to T or the arrow from H to Y is missing in Figure 6.1.

To investigate this, we first take a look at the statistical significance of the coefficients of H in the parametric exposure model, i.e. the model for the propensity score, via `prop.r` below. We see that all except that of `gt65` are statistically significant. Thus, the arrow from H to T is not missing.

```
prop.r <- function(data = gssrcc)
{
  # Fit the exposure model, i.e. the propensity score model
  out <-
```

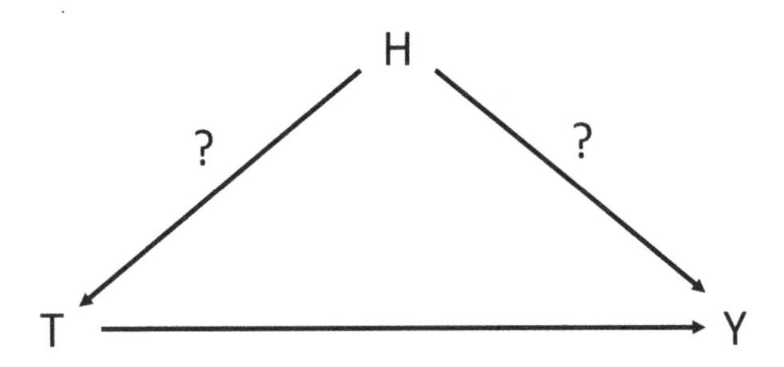

FIGURE 6.1: Is the Arrow from H to T or the Arrow from H to Y Missing?

```
    glm(gthsedu ~ magthsedu + white + female + gt65,
        family = binomial,
        data = data)
  # Return fitted values and a summary of the model fit
  list(e = fitted(out), emod = summary(out))
}
> prop.r()$emod
Call:
glm(formula = gthsedu ~ magthsedu + white + female + gt65, family
    = binomial, data = data)
Coefficients:
            Estimate Std. Error z value Pr(>|z|)
(Intercept) -1.20175    0.11195  -10.73  < 2e-16
magthsedu    1.29855    0.10684   12.15  < 2e-16
white        0.50066    0.10575    4.73  2.2e-06
female       0.19367    0.09248    2.09    0.036
gt65         0.00903    0.11384    0.08    0.937
```

Second, we take a look at the statistical significance of the coefficients of H in a parametric model for $E(Y|H)$, denoted by $d(H)$, which we will call the *prognostic score*, via `prog.r` below. Previously, Hansen (2008) defined a prognostic score as $E(Y|H, T = 0)$, under certain assumptions. Our prognostic score does not condition on $T = 0$, and it also differs from the outcome model by exclusion of T. We see that all of the coefficients of the model for $E(Y|H)$ are statistically significant. Thus, the arrow from H to Y is not missing.

```
prog.r <- function(data = gssrcc)
{
  # Fit the prognostic score model
  out <-
    glm(trump ~ magthsedu + white + female + gt65,
        family = binomial,
        data = data)
```

```
  # Return the fitted values and a summary of the model fit
  list(d = fitted(out), dmod = summary(out))
}
> prog.r()$dmod
Call:
glm(formula = trump ~ magthsedu + white + female + gt65,
    family = binomial, data = data)
Coefficients:
            Estimate Std. Error z value Pr(>|z|)
(Intercept)   -2.856      0.210  -13.60  < 2e-16
magthsedu     -0.499      0.132   -3.78  0.00016
white          2.374      0.209   11.37  < 2e-16
female        -0.503      0.106   -4.74  2.2e-06
gt65           0.571      0.121    4.73  2.3e-06
```

What is going on? How can there be no confounding of the effect of T on Y, at the same time as an arrow from H to T and an arrow from H to Y? The answer is that faithfulness must be violated; the arrows from H to T and from H to Y must cancel each other out. How do we check this? We turn to the causal DAG of Figure 6.2.

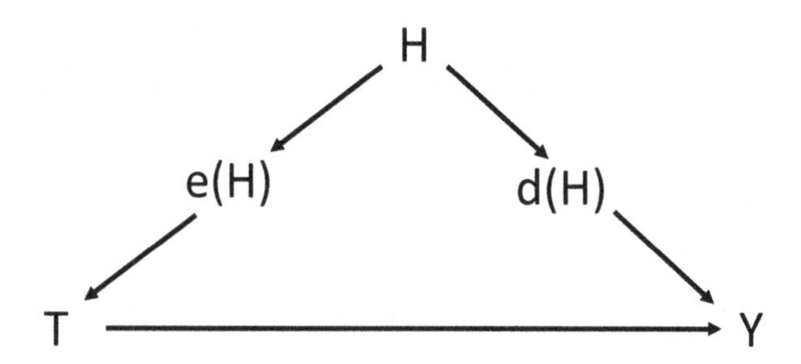

FIGURE 6.2: Violation of Faithfulness: The Arrows from H to $e(H)$ and from H to $d(H)$ Cancel Each Other in the Trump Example

Conveniently, both $e(H)$ and $d(H)$ are univariate, which enables us to measure their association by way of Pearson's correlation. If the correlation is essentially zero, then the arrows from H to $e(H)$ and from H to $d(H)$ are effectively canceling each other out. We check this with `cor.test`:

```
> cor.test(prop.r()$e, prog.r()$d)
        Pearson's product-moment correlation
data:  prop.r()$e and prog.r()$d
t = 0.866, df = 2170, p-value = 0.39
alternative hypothesis: true correlation is not equal to 0
```

```
95 percent confidence interval:
 -0.023514  0.060653
sample estimates:
      cor
0.018603
```

We see that the estimated correlation is near zero, at 0.019, and that the 95% confidence interval (−0.024, 0.061) is tight about zero. The computation of the confidence interval ignores that the computed $e(H)$ and $d(H)$ are estimated rather than known functions of H, but we have enough evidence to reject faithfulness, because the bootstrap confidence interval would not be much wider, and thus only very small values of the correlation are plausible. This example serves as a caution against assuming faithfulness too readily.

6.3 Doubly Robust Standardization

Validity of standardization with a parametric outcome or exposure model requires correctness of the chosen model. The two methods can yield quite different results, as observed in Zheng et al. (2013). Doubly robust standardization avoids this problem by using an estimator that relies on both models and that is valid if at least one of the models is correct (Bang and Robins (2005)).

The method is based on the following relations, that are true assuming (6.1), consistency, positivity, and that either the exposure model $e(H)$ or the outcome model $E(Y|H,T)$ is correctly specified.

$$E(Y(1)) = E\left(\frac{TY}{e(H)} - \frac{T-e(H)}{e(H)}E(Y|H,T=1)\right), \tag{6.17}$$

and

$$E(Y(0)) = E\left(\frac{(1-T)Y}{1-e(H)} + \frac{T-e(H)}{1-e(H)}E(Y|H,T=0)\right). \tag{6.18}$$

The proof of (6.17) follows. First, we assume that $e(H)$ is correct. In that case,

$$E\left(\frac{TY}{e(H)}\right) = E(Y(1)),$$

as proved for (6.9). Furthermore,

$$E\left(\frac{T-e(H)}{e(H)}E(Y|H,T=1)|H\right) = E\left(\frac{T-e(H)}{e(H)}|H\right)E(Y|H,T=1), \tag{6.19}$$

because $E(Y|H,T=1)$ is constant conditional on H. Then

$$E\left(\frac{T-e(H)}{e(H)}|H\right)=0,$$

because expectation is linear, $e(H)$ is constant conditional on H, and $E(T|H)=e(H)$. By the double expectation theorem,

$$E\left(\frac{T-e(H)}{e(H)}E(Y|H,T=1)\right)=0.$$

Thus, we have proved (6.17).

Second, we assume that $E(Y|H,T=1)$ is correct. We can rearrange the right hand side of (6.17) as

$$E(Y|H,T=1)+\frac{T}{e(H)}(Y-E(Y|H,T=1)).$$

The expectation of the first term is

$$E_H(E(Y|H,T=1)=E(Y(1)),$$

as proved for (6.2). We first take the conditional expectation of the second term given H and $T=1$, to find that it equals zero:

$$E\left(\frac{T}{e(H)}(Y-E(Y|H,T=1))|H,T=1\right)=$$
$$\frac{1}{e(H)}\left(E(Y|H,T=1)-E(Y|H,T=1)\right)=0.$$

Hence, the unconditional expectation of the second term also equals zero, by the double expectation theorem. Thus, we have proved (6.17). The proof of (6.18) is analogous.

Using (6.17) and (6.18), we can estimate $E(Y(1))$ and $E(Y(0))$ with

$$\hat{E}(Y(1))=(1/n)\left\{\Sigma_i\left(\frac{T_i}{\hat{e}(H_i)}Y_i-\frac{T_i-\hat{e}(H_i)}{\hat{e}(H_i)}\hat{E}(Y_i|H_i,T_i=1)\right)\right\} \quad (6.20)$$

and an analogous expression for $\hat{E}(Y(0))$.

To illustrate, we apply the method to the What-If? Study data assuming `lvlcont0` is a sufficient confounder for the effect of `A` on `vl4`. We use a misspecified outcome model including only `A` and the intercept, and we use an exposure model including the intercept and `lvlcont0`. The results of standardization using only the exposure model were shown previously in Table 6.9, and the results of standardization using only the misspecified outcome model are shown below in Table 6.11.

TABLE 6.11
Outcome-model Standardization for the What-If? Study with the Misspecified Outcome Model

Measure	Estimate	95% CI
$\hat{E}(Y(0))$	0.400	(0.277, 0.523)
$\hat{E}(Y(1))$	0.276	(0.190, 0.362)
$\hat{RD}$	−0.124	(−0.273, 0.025)
$\hat{RR}$	0.690	(0.442, 1.07)

Focusing on the risk difference, the estimate and 95% confidence interval are −0.060 (−0.188, 0.069) using the exposure model as compared to −0.124 (−0.273, 0.025) using the misspecified outcome model. Although neither method leads to statistical significance, because both confidence intervals include zero, the estimate using the misspecified outcome model suggests a larger effect that is close to statistical significance.

We use the R code below for estimation according to (6.20):

```
badstanddr.r <- function(data, ids)
{
  dat <- data[ids, ]
  # Fit the parametric exposure model
  e <- fitted(glm(A ~ lvlcont0, family = binomial, data = dat))
  # Fit a nonparametric outcome model that we do not believe
  # i.e., a bad outcome model
  lmod <- glm(vl4 ~ A, family = binomial, data = dat)
  dat0 <- dat1 <- dat
  dat0$A <- 0
  dat1$A <- 1
  # Predict potential outcomes for each participant
  EYhat0 <- predict(lmod, newdata = dat0, type = "response")
  EYhat1 <- predict(lmod, newdata = dat1, type = "response")
  # Use the DR estimating equation to estimate expected
  #  potential outcomes
  EY0 <- mean(dat$vl4 * (1 - dat$A) / (1 - e) + EYhat0 * (e - dat$A)
                                                          / (1 - e))
  EY1 <- mean(dat$vl4 * (dat$A / e) - EYhat1 * (dat$A - e) / e)
  # Estimate the effect measures
  rd <- EY1 - EY0
  rr <- log(EY1 / EY0)
  c(EY0, EY1, rd, rr)
}
```

The results are shown in Table 6.12. The estimate and 95% confidence interval for the risk difference are −0.062 (−0.183, 0.060), quite close to those of the exposure-modeling approach, which are themselves almost identical to those estimated with the outcome model using both `A` and `lvlcont0`, previously presented in Table 6.7. This should come as no surprise, as we saw previously that the conditional expectation of the outcome given `A` and `lvlcont0` depends on `lvlcont0`.

TABLE 6.12
Doubly Robust Standardization for the What-If? Study Combining the Misspecified Outcome Model of Table 6.11 and the Exposure Model of Table 6.9

Measure	Estimate	95% CI
$\hat{E}(Y(0))$	0.362	(0.253, 0.471)
$\hat{E}(Y(1))$	0.300	(0.216, 0.385)
$\hat{RD}$	−0.062	(−0.183, 0.060)
$\hat{RR}$	0.830	(0.571, 1.20)

It is natural to question when one should consider using standardization with an outcome model versus with an exposure model versus doubly robust standardization. When the first two methods give discrepant answers, doubly robust standardization might help choose between them. However, if we choose doubly robust standardization in the first place, we might pay a price in terms of variability of the estimate. One might think that standardization with the exposure model would be preferable when the outcome indicates a rare condition. To see this, first suppose the condition is not rare. We might have 3000 individuals and 50%, or 1500, with the condition. Using the rule of thumb for logistic regression of Peduzzi et al. (1996) presented in Chapter 2, we should be able to include 150 covariates in the outcome model. Now suppose the condition is rare, and 1%, or 30, have it. Our rule of thumb now suggests we can only include 3 covariates in the outcome model. If our sufficient confounder, H, is high-dimensional, what do we do? Suppose the exposure, T, is divided more evenly: that is, we have 600 with $T = 1$. This would suggest we can include 60 covariates in the exposure model. Would it not be preferable to use only the exposure model for standardization?

To investigate, we turned to the simulation study `simdr.r` with `ss` indicating the number of columns of H. The columns of H were independent indicator variables each with probability 0.05. We simulated T as indicator variables with probabilities that varied as a linear function of H, such that approximately 600 individuals had $T = 1$. We simulated Y_i as a function of T_i and $\Sigma_{k=1}^{\texttt{ss}} H_{ik}$, such that approximately 35 individuals had $Y = 1$. The mean of $\Sigma_{k=1}^{\texttt{ss}} H_{ik}$ was fixed at one, but when `ss` was set to 100, it ranged from 0.00 to 2.80. $P(T = 1|H)$ ranged from 0.041 to 0.468. $E(Y|T, H)$ ranged from 0.000 to 0.036. The range of $\hat{e}(H_i)$ fitted with a correctly specified linear exposure model was −0.037 to 0.575. Because the propensity score should not be negative, we also tried a logistic exposure model for exposure model standardization, even though it was not correctly specified. We let `ss`, which directly relates to the number of covariates in the model, equal 40 and 100. We would have expected exposure-model standardization to be best with 40, and we would have expected all methods to break down with 100. However, this did not happen. To try to force the outcome-modeling and doubly robust standardization to break down, we overspecified the outcome model, using

```
Y ~ T * H
```

to include all two-way interactions between T and the columns of H.

```
simdr.r <- function(ss = 100)
{
  # ss is the number of confounders
  # i.e. the number of columns of H
  H <- matrix(0, 3000, ss)
  # Let all components of H be independent Bernoulli with p=0.05
  probH <- rep(0.05, 3000)
  for (i in 1:ss)
  {
    H[, i] <- rbinom(n = 3000, size = 1, prob = probH)
  }
  # Let the treatment depend on a function of H
  sumH <- apply(H, 1, sum) * (20 / ss)
  # Need to make sure P(T=1) is between 0 and 1
  #return(range(sumH))
  probT <- .13 * sumH + .05 * rnorm(n = 3000, mean = 1, sd = .1)
  #return(range(probT))
  T <- rbinom(n = 3000, size = 1, prob = probT)
  #return(sum(T))
  # Generate the outcome depend on T and H
  probY <- .01 * T + .01 * sumH
  # Make sure P(Y=1) is between 0 and 1
  #return(range(probY))
  Y <- rbinom(n = 3000, size = 1, prob = probY)
  #return(sum(Y))
  # Fit the exposure model
  e <- fitted(lm(T ~ H))
  # Refit the exposure model using an incorrect logistic model
  e2 <- predict(glm(T ~ H, family = binomial), type = "response")
  # Explore the range of the propensity scores
  #return(range(e))
  # Compute the weights
  w0 <- (1 - T) / (1 - e)
  w1 <- T / e
  w02 <- (1 - T) / (1 - e2)
  w12 <- T / e2
  #return(summary(lm(Y~T*H)))
  dat0 <- dat1 <- dat <- data.frame(cbind(Y, T))
```

```
  dat0$T <- 0
  dat1$T <- 1
  # Fit an overspecified outcome model
  out <- lm(Y ~ T * H)
  # Estimate the expected potential outcomes using the various methods
  EY0out <- mean(predict(out, newdata = dat0))
  EY1out <- mean(predict(out, newdata = dat1))
  EY0exp <- weighted.mean(Y, w = w0)
  EY1exp <- weighted.mean(Y, w = w1)
  EY0exp2 <- weighted.mean(Y, w = w02)
  EY1exp2 <- weighted.mean(Y, w = w12)
  EY0dr <- mean(w0 * Y + predict(out, newdata = dat0) * (T - e)
                                                      / (1 - e))
  EY1dr <- mean(w1 * Y - predict(out, newdata = dat1) * (T - e) / e)
  EYT0 = mean(Y * (1 - T))
  EYT1 = mean(Y * T)
  list(
    EY0exp = EY0exp,
    EY1exp = EY1exp,
    EY0exp2 = EY0exp2,
    EY1exp2 = EY1exp2,
    EYT0 = EYT0,
    EYT1 = EYT1,
    EY0out = EY0out,
    EY1out = EY1out,
    EY0dr = EY0dr,
    EY1dr = EY1dr
  )
}
```

The results of the simulation study that collected the results of running `simdr.r` 1000 times, in order to obtain the sampling distribution of the estimators, are presented in Tables 6.13 and 6.14. The unadjusted estimator `EYT0` estimates $E(Y|T = 0)$, and `EYT1` estimates $E(Y|T = 1)$. The estimators `EY0exp` and `EY1exp` use exposure model standardization with the correctly specified linear exposure model, whereas `EY0exp2` and `EY1exp2` use exposure model standardization with the incorrect logistic exposure model. The estimators `EY0out` and `EY1out` use outcome model standardization with the overspecified outcome model. The doubly robust estimators `EY0dr` and `EY1dr` use the overspecified outcome model and the correctly specified linear exposure model. The mean and standard deviation columns show the means and standard deviations of the sampling distributions of the estimators. The P-value column tests whether our sample of 1000 estimators comes from a sampling distribution with a true mean equal to $E(Y(0)) = 0.01$ or $E(Y(1)) = 0.02$, whichever is relevant.

The only estimators with P-values < 0.05 are the unadjusted estimators `EYT0` and `EYT1`. Indeed, we observe from the means of their sampling distribution, 0.0076 and 0.0042, that the unadjusted estimators are biased for the true values, 0.01 and 0.02. This confirms that standardization is necessary. The standard deviations of the sampling distribution indicate how variable the estimators are; smaller standard deviations are desirable, provided bias is negligible. We see that the standard deviation of the sampling distribution is larger for the correctly specified linear exposure model than it is for the incorrectly specified logistic model, which happens to exhibit negligible bias. Surprisingly, the standard deviations for the overspecified outcome model are about the same as for the logistic exposure model, despite that the outcome is rare. Previously, Vittinghoff and McCulloch (2006) investigated the performance of the outcome model for conditional effect estimates and also found it to be quite robust to inclusion of many confounders. Finally, we see large standard deviations for the doubly robust estimators of $E(Y(1))$, particularly for 100 columns of H. The doubly robust estimator appears to be the first to break down as we move from 40 to 100 confounders, with the correctly specified linear exposure model second. In conclusion, all of the methods were robust with 40 confounders, and even with 100 confounders, results were not horrible. In particular, we question the logic that might lead to a choice of exposure-model standardization over outcome-model standardization with the outcome indicating a rare condition. Furthermore, we caution against using the doubly robust approach with high-dimensional confounders.

TABLE 6.13
Sampling Distribution of Estimators from Simulation Study Investigating Small-Sample Robustness: True $E(Y(0)) = 0.01$, True $E(Y(1)) = 0.02$, with 40 Columns of H

Estimator	Description	Mean	Standard Deviation	P-Value
`EYT0`	Unadjusted	0.0076	0.0015	0.00
`EYT1`	Unadjusted	0.0042	0.0012	0.00
`EY0exp`	Linear Exposure Model	0.0100	0.0021	0.92
`EY1exp`	Linear Exposure Model	0.0195	0.0127	0.19
`EY0exp2`	Logistic Exposure Model	0.0101	0.0021	0.42
`EY1exp2`	Logistic Exposure Model	0.0204	0.0064	0.07
`EY0out`	Overspecified Outcome Model	0.0100	0.0021	0.79
`EY1out`	Overspecified Outcome Model	0.0200	0.0066	0.84
`EY0dr`	Doubly Robust	0.0100	0.0021	0.82
`EY1dr`	Douby Robust	0.0197	0.0106	0.37

TABLE 6.14
Estimated Sampling Distribution of Estimators from Simulation Study Investigating Small-Sample Robustness: True $E(Y(0)) = 0.01$, True $E(Y(1)) = 0.02$, with 100 Columns of H

Estimator	Description	Mean	Standard Deviation	P-Value
`EYT0`	Unadjusted	0.0079	0.0016	0.00
`EYT1`	Unadjusted	0.0038	0.0012	0.00
`EY0exp`	Linear Exposure Model	0.0100	0.0020	0.61
`EY1exp`	Linear Exposure Model	0.0196	0.0562	0.81
`EY0exp2`	Logistic Exposure Model	0.0100	0.0020	0.73
`EY1exp2`	Logistic Exposure Model	0.0200	0.0068	0.96
`EY0out`	Overspecified Outcome Model	0.0100	0.0020	0.74
`EY1out`	Overspecified Outcome Model	0.0200	0.0069	0.74
`EY0dr`	Doubly Robust	0.0100	0.0020	0.72
`EY1dr`	Douby Robust	0.029	0.1891	0.14

6.4 Exercises

1. Subset the `brfss` data introduced in Chapter 3 to include only those with `gt65` equal to zero. Apply the outcome-modeling approach to standardization to estimate the effect of `insured` on `flushot`, adjusting for the confounders `female`, `whitenh`, `blacknh`, `hisp`, `multinh`, `gthsedu`, and `rural`. Estimate the potential outcomes, the risk difference, the relative risk, and the odds ratio. Use the bootstrap to construct confidence intervals. Because the dataset is so large, try using only 500 bootstrap samples. Interpret your results, being careful to state any assumptions required for validity.
2. Repeat the first exercise but using the exposure-modeling approach to standardization.
3. Repeat the first exercise but using the doubly robust approach to standardization.
4. Refer back to the two-part model estimated from the `brfss` data in exercise 3 of Chapter 3. In this exercise, we will piece together the two parts of the model to estimate the expected largest number of alcoholic drinks on any occasion unconditional on whether alcohol has been consumed or not. Let the confounders be summarized by H and include `gt65`, `female`, `whitenh`, `blacknh`, `hisp`, `multinh`, and `gthsedu`. Denote the exposure by R, for `rural`. Let Y denote the outcome, `maxdrinks`, and let Z denote `zerodrinks`. Note that $Z = 0$ exactly when $Y > 0$. The logistic parametric model is a model for $P(Z = 1|R, H)$, and the loglinear parametric model is a model for $E(Y|Z = 0, R, H)$. Using the double expectation theorem, we

have that

$$\begin{aligned} E(Y|R,H) &= E(Y|Z=1,R,H)P(Z=1|R,H) \\ &+ E(Y|Z=0,R,H)(1-P(Z=1|R,H)), \end{aligned}$$

which reduces to

$$E(Y|R,H) = E(Y|Z=0,R,H)(1-P(Z=1|R,H)), \qquad (6.21)$$

because $E(Y|Z=1,R,H)=0$. The two-part model is particularly useful when the outcome Y is frequently zero, in part because a loglinear model for Y would never predict a zero outcome.

Use the result at (6.21) in conjunction with the outcome-modeling approach to standardization to estimate

$$E_H E(Y|R=r,H) = E_H\left[E(Y|Z=0,R=r,H)(1-P(Z=1|R=r,H))\right]$$

for $r=0,1$. Contrast the results for $r=0,1$ using a difference and a ratio. Use the bootstrap to construct confidence intervals. Interpret your results, being careful to state any assumptions required for validity.

Note that once the two pieces of the two-part model are estimated and used to predict $E(Y|Z=0,R=r,H)$ and $P(Z=1|R=r,H)$, the outer expectation E_H is just a sample average. Be sure to predict $E(Y|Z=0,R=r,H)$ for *everyone* in the sample, and not just those with $Z=0$.

5. Compare the results of the previous exercise with those obtained from an exposure-modeling approach to standardization. Note that with the exposure-modeling approach, there is no need to use a two-part model.
6. Combine the outcome-modeling and exposure-modeling approaches of the previous two exercises to construct a doubly robust estimator, and again contrast the results for $r=0,1$ using a difference and a ratio. Use the bootstrap to construct confidence intervals. Compare your results with those from the other two approaches.
7. Estimate the ATT using the exposure-modeling approach for comparison with the effect estimated in exercise 5.
8. Derive a relation we can use to estimate $E(Y(1)|T=0)$ using an exposure-modeling approach.

7

Adjusting for Confounding: Difference-in-Differences Estimators

On March 11, 2020, the World Health Organization (WHO) declared COVID-19 a pandemic. The public health crisis affected countries across the globe, with several, including the US, shutting down large segments of their economies in order to stem the spread of the coronavirus. In the US, estimated monthly employment rates for men and women from January to August 2020 are reported in Table 7.1. These estimates were produced by the US Bureau of Labor Statistics using data from the Current Population Survey, which is a monthly survey of households conducted by the Bureau of Census. The estimates are seasonally adjusted, meaning that increases or decreases from month to month are due to factors other than seasonal variation for a typical year. These data document, in a very obvious way, that shutting down the economy subsequent to the pandemic declaration triggered a large rise in unemployment. We see a relatively small rise from February to March, followed by a large rise in April and beyond. In this example, no one would question that the increase was caused by the nation's response to the pandemic. In the face of cause and effect this obvious, sophisticated statistical methods are not really needed. Simple subtraction, i.e. 16.2% – 4.4% = 11.8% for women or 13.5% – 4.4% = 9.1% for men, is enough.

TABLE 7.1
US Monthly Percent Unemployment from January to August 2020 of the Civilian Noninstitutional Population 16 Years and Over, Estimated by the Bureau of Labor Statistics Using Current Population Survey Data

Month	Percent of Men	Percent of Women
January	3.6	3.5
February	3.6	3.4
March	4.4	4.4
April	13.5	16.2
May	12.2	14.5
June	10.6	11.7
July	9.8	10.6
August	8.3	8.6

DOI: 10.1201/9781003146674-7

In this example, we were able to rule out seasonal variation as a major confounder for two reasons. First, the estimates were reported as seasonally adjusted, meaning that this confounder has already been removed, perhaps via standardization. Second, the magnitude of the differences is too great to be due to seasonal fluctuations alone. For many other investigations involving pre- versus post-exposure differences, cause and effect is not as obvious.

For example, Molyneux et al. (2019) estimated the effect of negative interest rate policy (NIRP) on bank margins and profits. Responding to the global financial crisis of 2007–2008, the central banks of many countries implemented NIRP in order to provide economic stimulus to weak economies. Former President Trump repeatedly tweeted about the benefits of negative interest rates. For example, on September 3, 2019, @realDonaldTrump tweeted "Germany, and so many other countries, have negative interest rates, 'they get paid for loaning money,' and our Federal Reserve fails to act! Remember, these are also our weak currency competitors!" Many economists are not as sanguine. Molyneux et al. (2019) analyzed a dataset comprising 7,359 banks from 33 OECD member countries over 2012–2016 to assess the impact of NIRP on net interest margins (NIMs). NIM measures the net amount a bank earns on loans and other interest-earning assets relative to the amount of those loans and other assets. For example, supposing the bank's interest earning assets equal one million dollars in a year, the bank earned \$50,000 in interest at 5% and paid \$20,000 in expenses to their lenders, then the NIM would be (\$50,000 − \$20,000)/\$1,000,000 = 3%. Table 2 of Molyneux et al. (2019) reports on yearly NIMs of banks in countries initiating NIRP both pre and post-NIRP. There are 8916 bank-years (one bank-year represents one NIM from one bank from one year) pre-NIRP with an average NIM of 2.06% and a standard deviation of 0.95% (hence a standard error of $0.95/\sqrt{8916} = 0.0100\%$) and 8040 bank-years post-NIRP with an average NIM of 1.92% and a standard deviation of 0.78% (hence a standard error of $0.78/\sqrt{8040} = 0.0087\%$). Ignoring the temporal correlation between NIMs of a single bank from year to year, we can assess whether the difference in average NIMs, that is $1.92\% - 2.06\% = -0.14\%$, is statistically significant using a z-test, computing

$$z = \frac{-0.14}{\sqrt{0.0100^2 + 0.0087^2}} = -10.562,$$

which indicates that the difference is highly statistically significant. However, it may be due to factors other than initiation of NIRP. Perhaps temporal changes in other variables led to the difference over time.

The difference-in-differences solution incorporates a control group, resulting in one of the most popular tools for applied research in economics to evaluate the effects of public interventions and other treatments of interest on relevant outcome variables (Abadie (2005)). We compare the change over time in the exposed group to the change over time in the unexposed group.

Molyneux et al. (2019) also present statistics on banks in countries that did not initiate NIRP, over a matched time period. There are 4686 control bank-years pre-NIRP with an average NIM of 2.92% and a standard deviation of 1.71% (hence a standard error of $1.71/\sqrt{4686} = 0.0250\%$) and 4331 control bank-years post-NIRP with an average NIM of 2.93% and a standard deviation of 1.65% (hence a standard error of $1.65/\sqrt{4331} = 0.0251\%$). We use the following R code to determine whether the difference in differences, that is, $(1.92 - 2.06) - (2.93 - 2.92) = -0.15\%$, is statistically significant.

```
analyze.r <- function ()
{
  # Compute the DiD Estimator
  did <- (2.06 - 1.92) - (2.92 - 2.93)
  # Estimate its standard error
  se <-
    sqrt((.95 ^ 2) / 8916 + (.78 ^ 2) / 8040 + (1.71 ^ 2) /
           4686 + (1.65 ^ 2) / 4331)
  # Return the estimator, SE, and z-score
  list(did = did, se = se, z = did / se)
}
> analyze.r()
$did
[1] 0.15
$se
[1] 0.037809
$z
[1] 3.9673
>2*pnorm(-3.9673)
[1] 7.2691e-05
```

The z-statistic is 3.97, corresponding to a P-value less than 0.0001, indicating that the difference in differences is indeed statistically significant, which suggests that initiation of NIRP negatively affects banks. In the next section, we provide the foundation for this approach to adjusting for confounding. We note that Molyneux et al. (2019) also applied a more complicated difference-in-differences approach that accounted for the temporal correlation of NIMs from a single bank and also adjusted for other factors, but their results were qualitatively the same as ours.

7.1 Difference-in-Differences (DiD) Estimators with Linear, Loglinear, and Logistic Models

Let Y_t, $t = 0, 1$, denote the pre- and post-exposure measures. Let A indicate the exposure. Let $Y_1(0)$ and $Y_1(1)$ denote the potential post-exposure outcomes to $A = 0$ and $A = 1$, respectively.

7.1.1 DiD Estimator with a Linear Model

The method relies on consistency as well as assumption A1:

$$\text{A1:}\;\; E(Y_1(0)|A=1) - E(Y_1(0)|A=0) = E(Y_0|A=1) - E(Y_0|A=0), \tag{7.1}$$

also called additive equi-confounding by Hernan and Robins (2020). The target of estimation is the linear ATT, presented in Chapter 6 as a risk difference, but also valid for non-binary Y_t:

$$\text{Linear ATT:}\;\; E(Y_1(1) - Y_1(0)|A=1). \tag{7.2}$$

The DiD estimator derives from the relation

$$\begin{aligned} E(Y_1(1) - Y_1(0)|A=1) = {} & E(Y_1|A=1) - E(Y_1|A=0) - \\ & (E(Y_0|A=1) - E(Y_0|A=0)), \end{aligned} \tag{7.3}$$

which connects the estimand framed in terms of potential outcomes to an estimand relying only on observed data. To prove the relation, note that

$$E(Y_1|A=1) - E(Y_1|A=0) = E(Y_1(1)|A=1) - E(Y_1(0)|A=0),$$

by consistency, and furthermore that

$$E(Y_0|A=1) - E(Y_0|A=0) = E(Y_1(0)|A=1) - E(Y_1(0)|A=0),$$

by assumption A1. Taking the difference of these two differences proves the relation.

We can therefore estimate the linear ATT at (7.2) via the difference in differences of averages,

$$\hat{E}(Y_1|A=1) - \hat{E}(Y_1|A=0) - (\hat{E}(Y_0|A=1) - \hat{E}(Y_0|A=0)),$$

which equals

$$\hat{E}(Y_1 - Y_0|A=1) - \hat{E}(Y_1 - Y_0|A=0). \tag{7.4}$$

This is the estimator we used for the NIRP example.

We can also compute the DiD estimator via the linear model

$$E(Y_t|A) = \alpha_0 + \alpha_1 t + \alpha_2 A + \beta A * t, \tag{7.5}$$

where

$$\begin{aligned} \beta = E(Y_1|A=1) - E(Y_1|A=0) - (E(Y_0|A=1) - E(Y_0|A=0)) \\ = (\alpha_0 + \alpha_1 + \alpha_2 + \beta) - (\alpha_0 + \alpha_1) \\ - ((\alpha_0 + \alpha_2) - \alpha_0)\,. \end{aligned}$$

Thus β is the linear ATT, and we can use linear regression to estimate it.

As we can estimate $E(Y(1)|A=1)$ directly via $E(Y_1|A=1)$, we can also recover $E(Y(0)|A=1)$ via $E(Y_1|A=1) - \beta$. Thus we can also estimate the loglinear and logistic ATTs using the linear model. Section 7.2 provides an example in `R`.

7.1.2 DiD Estimator with a Loglinear Model

The method relies on consistency as well as assumption A2:

$$\text{A2:}\quad E(Y_1(0)|A=1)/E(Y_1(0)|A=0) = E(Y_0|A=1)/E(Y_0|A=0), \tag{7.6}$$

which one might also call multiplicative equi-confounding or additive equi-confounding on the log scale. The target of estimation is the loglinear ATT:

$$E(Y_1(1)|A=1)/E(Y_1(0)|A=1). \tag{7.7}$$

The DiD estimator derives from the relation

$$\begin{aligned}\log E(Y_1(1)|A=1) - \log E(Y_1(0)|A=1) = \\ \log E(Y_1|A=1) - \log E(Y_1|A=0) - (\log E(Y_0|A=1) - \log E(Y_0|A=0)),\end{aligned}$$

which connects the estimand framed in terms of potential outcomes to an estimand relying only on observed data. The proof is similar to that for relation (7.3), relying on consistency and assumption A2.

We can therefore estimate the log of the loglinear ATT at (7.7) via the difference in differences of log averages, analogous to (7.4), and then exponentiate. We can also compute the DiD estimator via the loglinear model

$$\log E(Y_t|A) = \alpha_0 + \alpha_1 t + \alpha_2 A + \beta A * t, \tag{7.8}$$

where

$$\begin{aligned}\beta &= \log E(Y_1|A=1) - \log E(Y_1|A=0) - (\log E(Y_0|A=1) - \log E(Y_0|A=0)) \\ &= (\alpha_0 + \alpha_1 + \alpha_2 + \beta) - (\alpha_0 + \alpha_1) \\ &\quad - ((\alpha_0 + \alpha_2) - \alpha_0).\end{aligned}$$

Thus β is the log of the loglinear ATT, and we can use loglinear regression to estimate it. We then exponentiate the results. Again, as we can estimate $E(Y(1)|A=1)$ directly via $E(Y_1|A=1)$, we can also recover $E(Y(0)|A=1)$ via $E(Y_1|A=1)/\exp(\beta)$. Thus we can also estimate the linear and logistic ATTs using the loglinear model. Section 7.2 provides an example in R.

7.1.3 DiD Estimator with a Logistic Model

The method relies on consistency as well as assumption A3:

$$\begin{aligned}\text{A3:}\quad &\text{logit}E(Y_1(0)|A=1) - \text{logit}E(Y_1(0)|A=0) = \\ &\text{logit}E(Y_0|A=1) - \text{logit}E(Y_0|A=0),\end{aligned}$$

which one might also call additive equi-confounding on the logit scale. The target of estimation is the logistic ATT:

$$\text{logit}E(Y_1(1)|A=1) - \text{logit}E(Y_1(0)|A=1). \tag{7.9}$$

The DiD estimator derives from the relation

$$\text{logit}E(Y_1(1)|A=1) - \text{logit}E(Y_1(0)|A=1) =$$
$$\text{logit}E(Y_1|A=1) - \text{logit}E(Y_1|A=0) - (\text{logit}E(Y_0|A=1) - \text{logit}E(Y_0|A=0)),$$

which connects the estimand framed in terms of potential outcomes to an estimand relying only on observed data. The proof is similar to that for relation (7.3), relying on consistency and assumption A3.

We can therefore estimate the logistic ATT at (7.9) via the difference in differences of logit averages, analogous to (7.4). For binary Y_t, exponentiating yields an odds ratio. We can also compute the DiD estimator via the logistic model using the regression

$$\text{logit}E(Y_t|A) = \alpha_0 + \alpha_1 t + \alpha_2 A + \beta A * t, \tag{7.10}$$

where β is the logistic ATT, and we can use logistic regression to estimate it. Again, for binary Y_t, exponentiating yields an odds ratio. Again, as we can estimate $E(Y(1)|A=1)$ directly via $E(Y_1|A=1)$, we can also recover $E(Y(0)|A=1)$ by applying the expit function to $\text{logit}(E(Y_1|A=1)) - \beta$. Thus we can also estimate the linear and loglinear ATTs using the logistic model. Section 7.2 provides an example in R.

7.2 Comparison with Standardization

The DiD approach to adjusting for confounding relies on consistency plus one of assumptions A1, A2, or A3, which involve the confounder Y_0. We saw in Chapter 6 that we can also use standardization to estimate the linear, loglinear, or logistic ATT. However, standardization relies on consistency plus the sufficient confounder assumption (6.1). Interestingly, and as we will see below in the case of the Double What-If? Study, when Y_0 is not a sufficient confounder, it can still happen that one of assumptions A1, A2, or A3 is true. This implies that a DiD analysis may be valid even when standardization using only Y_0 may not be. Conversely, Y_0 may be a sufficient confounder without A1, A2, or A3 holding. In that case, standardization using only Y_0 would be valid whereas the DiD analyses would not be. It is worth noting that when one of A1, A2, or A3 holds, the other two will typically not hold. Therefore, the validity of the DiD approach depends on correctly choosing which one might plausibly hold, often an impossible task. However, in our analysis of the Double What-If? Study, we observe that the confidence intervals for all three estimators include the true value for their respective estimands.

In our exposition of the three DiD approaches, we assumed Y_0 was a pre-exposure version of the post-exposure Y_1. More generally, we could let Y_0 represent another variable, say H, that could reasonably be presumed to satisfy one of the assumptions A1, A2, or A3, and the corresponding DiD estimator would be plausibly valid. We refer the reader to Sofer et al. (2016).

For binary datasets, the estimand corresponding to the DiD estimator with a linear model can be expressed as

$$E(Y_1|A=1) - (E(Y_1|A=0,Y_0=1) - E(Y_1|A=0,Y_0=0))\,E(Y_0|A=0) \\ -E(Y_1|A=0,Y_0=0) - (E(Y_0|A=1) - E(Y_0|A=0))\,, \qquad (7.11)$$

by substituting

$$E(Y_1|A=0) = E(Y_1|A=0,Y_0=0)(1-E(Y_0|A=0)) \\ +E(Y_1|A=0,Y_0=1)E(Y_0|A=0)$$

into the relation at (7.3). On the other hand, the estimand corresponding to the standardized ATT can be expressed as

$$E(Y_1|A=1) - (E(Y_1|A=0,Y_0=1) - E(Y_1|A=0,Y_0=0))\,E(Y_0|A=1) \\ -E(Y_1|A=0,Y_0=0). \qquad (7.12)$$

Subtracting (7.12) from (7.11) yields

$$(E(Y_1|A=0,Y_0=1) - E(Y_1|A=0,Y_0=0) - 1)\,(E(Y_0|A=1) - E(Y_0|A=0))\,. \qquad (7.13)$$

Therefore, the two estimands will differ unless either (a) $Y_1 = Y_0$ for everyone with $A = 0$, or (b) $Y_0 \amalg A$. In the first case, the ATT equals $E(Y_1 - Y_0|A = 1)$, and DiD estimation is unnecessary. In the second case, the ATT equals $E(Y_1|A = 1) - E(Y_1|A = 0)$, and standardization is unnecessary. Because the estimands will typically differ, we recommend trying both approaches whenever possible, as it is unlikely that scientific considerations will be able to distinguish which approach has more validity in a given context.

Next we consider examples. Recall that the Double What-If? Study was simulated according to `doublewhatifsim.r` in Chapter 1. The pre- and post-exposure measures are VL_0 and VL_1, which take the place of Y_0 and Y_1 above. In Chapter 6, we used the `doublewhatifsim.r` program to derive $E(VL_1(0)|A = 1) = 0.559$ and $E(VL_1(1)|A = 1) = 0.199$.

From the code, we see that `VL1prob = VL0prob + .1 - .45*AD1`, `VL0prob = .8 - .4*AD0`, and `AD1prob = .1 + .8A`. We can write

$$VL_1 = 0.8 - 0.4 * AD_0 + 0.1 - 0.45 * (0.1) - 0.45 * 0.8 * A + \epsilon_1 \qquad (7.14)$$

and

$$VL_0 = 0.8 - 0.4 * AD_0 + \epsilon_0,$$

where ϵ_0 and ϵ_1 are independent of all of the other variables. Therefore,

$$E(VL_1 - VL_0|A=1) = 0.1 - 0.45 * (0.1) - 0.45 * 0.8$$

and

$$E(VL_1 - VL_0|A = 0) = 0.1 - 0.45 * (0.1),$$

so that

$$E(VL_1 - VL_0|A = 1) - E(VL_1 - VL_0|A = 0) = -0.45 * 0.8 = -0.36.$$

Therefore, the expected value of the DiD estimator with a linear model equals -0.36. From Chapter 6, we have that that the linear ATT also equals $E(VL_1(1)|A = 1) - E(VL_1(0)|A = 1) = -0.36$, which means that the DiD estimator with a linear model is valid in this example. We can derive the linear ATT in an easier way (because we do not need to estimate the components of the difference) as follows. We have

$$E(VL_1(1)|A = 1) = E(VL_1|A = 1) =$$
$$0.8 - 0.4 * E(AD_0|A = 1) + 0.1 - 0.45 * (0.1 + 0.8)$$

and

$$E(VL_1|A = 0, AD_0) = 0.8 - 0.4 * AD_0 + 0.1 - 0.45 * (0.1)$$

both from (7.14). Recall that AD_0 is a sufficient confounder. Therefore, using standardization,

$$E(VL_1(0)|A = 1) = E_{AD_0|A=1}E(VL_1|A = 0, AD_0) =$$
$$0.8 - 0.4 * E(AD_0|A = 1) + 0.1 - 0.45 * 0.1,$$

so that

$$E(VL_1(1)|A = 1) - E(VL_1(0)|A = 1) = -0.45 * 0.8 = -0.36.$$

It is not likely that the DiD estimators with the loglinear model and the logistic model are valid, but we present these for the sake of comparison.

We next present the R code we used to compute the DiD estimators, which relied on the regression model formulations of (7.5), (7.8), and (7.10). First, we needed to transform the doublewhatifdat dataset from *short form* into *long form*. Long form contains one row per person per time period, whereas short form contains one row per person, which includes measures at both time periods. We do that as follows.

```
mklong.r <- function(dat = doublewhatifdat)
{
  # Set up the dataset
  longdat <-
    data.frame(
      "Y" = rep(0, 2 * nrow(dat)),
      "A" = rep(dat[, "A"], each = 2),
      "time" = rep(c(0, 1), times = nrow(dat))
    )
  # Insert Y_0
```

```
  longdat$Y[c(TRUE, FALSE)] <- dat[, "VL0"]
  # Insert Y_1
  longdat$Y[c(FALSE, TRUE)] <- dat[, "VL1"]
  longdat
}
> longdat <- mklong.r()
> head(longdat)
  Y A time
1 1 0    0
2 1 0    1
3 1 0    0
4 1 0    1
5 1 0    0
6 1 0    1
```

Next we compute β and our other measures using `didlinear.r`, `didloglinear.r`, and `didlogistic.r`, and we estimate confidence intervals using `bootdid.r`. We need to bootstrap `doublewhatifdat` rather than `longdat` so that we take a bootstrap sample of the participants rather than of the time points. For that reason, we call `mklong.r` from within `didlinear.r`, `didloglinear.r`, and `didlogistic.r`.

```
didlinear.r <-
  function(data = doublewhatifdat,
           ids = c(1:nrow(doublewhatifdat)))
  {
    dat <- data[ids, ]
    # Make the long form dataset
    dat <- mklong.r(dat)
    # Fit the linear DiD model
    beta <- lm(Y ~ A + time + A * time, data = dat)$coef
    # Extract the risk difference
    rd <- beta[4]
    # Estimate E(Y(1)|A=1)
    EY1 <- mean(dat$Y[(dat$A == 1) & (dat$time == 1)])
    # Estimate E(Y(0)|A=1)
    EY0 <- EY1 - rd
    # Estimate the other effect measures
    logrr <- log(EY1) - log(EY0)
    logor <- log(EY1) - log(1 - EY1) - log(EY0) + log(1 - EY0)
    c(EY0, EY1, rd, logrr, logor)
  }
didloglinear.r <-
  function(data = doublewhatifdat,
           ids = c(1:nrow(doublewhatifdat)))
  {
    dat <- data[ids, ]
    #Make the long form dataset
    dat <- mklong.r(dat)
    # Fit the loglinear DiD model
    beta <- glm(Y ~ A + time + A * time, family = poisson, data = dat)$coef
```

```
    # Extract the log relative risk
    logrr <- beta[4]
    # Estimate E(Y(1)|A=1)
    EY1 <- mean(dat$Y[(dat$A == 1) & (dat$time == 1)])
    # Estimate E(Y(0)|A=1)
    EY0 <- EY1 / exp(logrr)
    # Estimate the other effect measures
    rd <- EY1 - EY0
    logor <- log(EY1) - log(1 - EY1) - log(EY0) + log(1 - EY0)
    c(EY0, EY1, rd, logrr, logor)
  }
didlogistic.r <-
  function(data = doublewhatifdat,
           ids = c(1:nrow(doublewhatifdat)))
  {
    dat <- data[ids, ]
    # Make the long form dataset
    dat <- mklong.r(dat)
    # Fit the logistic DiD model
    beta <- glm(Y ~ A + time + A * time, family = binomial,
        data = dat)$coef
    # Extract the log odds ratio
    logor <- beta[4]
    # Estimate E(Y(1)|A=1)
    EY1 <- mean(dat$Y[(dat$A == 1) & (dat$time == 1)])
    # Estimate E(Y(0)|A=1)
    tmp <- log(EY1 / (1 - EY1)) - logor
    EY0 <- exp(tmp) / (1 + exp(tmp))
    # Estimate the other effect measures
    rd <- EY1 - EY0
    logrr <- log(EY1) - log(EY0)
    c(EY0, EY1, rd, logrr, logor)\enlargethispage{12pt}
  }
bootdid.r <- function ()
{
  stand.out <- boot(data = doublewhatifdat,
                    statistic = didlogistic.r,
                    R = 1000)
  stand.est <- summary(stand.out)$original
  stand.SE <- summary(stand.out)$bootSE
  stand.lci <- stand.est - 1.96 * stand.SE
  stand.uci <- stand.est + 1.96 * stand.SE
  list(
    stand.est = stand.est,
    stand.SE = stand.SE,
    stand.lci = stand.lci,
    stand.uci = stand.uci
  )
}
```

TABLE 7.2
Difference-in-Differences Estimation of the ATT for the Double What-If? Study

Method	Measure	Truth	Estimate	95% CI
All	$E(VL_1 \mid A=1)$	0.199	0.231	(0.179, 0.282)
Linear	$E(Y(0) \mid A=1)$	0.559	0.586	(0.508, 0.664)
Loglinear	$E(Y(0) \mid A=1)$	0.559	0.577	(0.498, 0.656)
Logistic	$E(Y(0) \mid A=1)$	0.559	0.592	(0.513, 0.671)
Linear	RD	−0.360	−0.355	(−0.441, −0.270)
Loglinear	RD	−0.360	−0.346	(−0.431, −0.262)
Logistic	RD	−0.360	−0.362	(−0.447, −0.276)
Linear	RR	0.356	0.394	(0.309, 0.5)
Loglinear	RR	0.356	0.400	(0.315, 0.508)
Logistic	RR	0.356	0.390	(0.306, 0.496)
Linear	OR	0.196	0.212	(0.142, 0.315)
Loglinear	OR	0.196	0.220	(0.149, 0.325)
Logistic	OR	0.196	0.206	(0.139, 0.307)

We observe that the three confidence intervals each contains the true value for their respective measures. We expected this for RD, but we did not necessarily expect it for RR and OR, as `doublewhatifsim.r` was not designed for assumptions A2 and A3 to hold. It is of interest to compare these results to those presented in Table 6.5, which reported the standardized estimates correctly using AD_0 as the sufficient confounder; $\hat{RD}$ was estimated at −0.344 (−0.404, −0.283) and $\hat{RR}$ was estimated at 0.402 (0.322, 0.501). It is also of interest to compare to Table 6.6, which reported the standardized estimates incorrectly using VL_0 as the sufficient confouder; $\hat{RD}$ was estimated with bias at −0.452 (−0.514, −0.389) and $\hat{RR}$ was estimated at 0.338 (0.267, 0.428).

For one last example, we apply DiD estimation to analyze the ATT for the What-If? Study, for comparison with Table 6.4, where $\hat{RD}$ was estimated at −0.085 (−0.207, 0.037) and $\hat{RR}$ was estimated at 0.765 (0.520, 1.12). For simplicity, we report just the risk difference for the linear method, the relative risk for the loglinear method, and the odds ratio for the logistic method.

TABLE 7.3
Difference-in-Differences Estimation of the ATT for the What-If? Study

Method	Measure	Estimate	95% CI
Linear	RD	−0.057	(−0.198, 0.083)
Loglinear	RR	0.829	(0.551, 1.26)
Logistic	OR	0.763	(0.408, 1.43)

We observe that the DiD point estimates are closer to the null than those obtained by standardization, but that the qualitative conclusion remains unchanged; i.e., the effect is not statistically significant.

In closing, DiD estimation is an attractive method to adjust for confounding in settings involving comparisons of pre- and post-exposure measures. However, standardization can also be used in those settings. We recommend trying both approaches, to make sure that results are qualitatively similar. If not, then caution is warranted in interpreting results, and the study may be deemed inconclusive.

7.3 Exercises

1. The `sepsisb` binary dataset, related to the `sepsis` binary dataset from Exercise 2 of Chapter 4, is taken from a prospective study conducted by the University of Florida Sepsis and Critical Illness Research Center; see Loftus et al. (2017) for the study design. All patients were septic at study entry. The variables are `Zubrodbase` indicating a Zubrod score of 3 or 4 at baseline, i.e. prior to hospitalization with sepsis, `shock` indicating presence of septic shock at study entry, and `Zubrod1yr` indicating a Zubrod score of 4 or 5 one year after study entry. A Zubrod score of 3 reflects that the participant is capable of only limited self-care and is confined to a bed or chair for more than 50% of waking hours. A Zubrod score of 4 reflects that the participant is confined to bed at all times, and a score of 5 reflects that the participant is dead. We observe that the baseline dichotomy differs from the one-year dichotomy of Zubrod by including scores of 3 and not including scores of 5. Nevertheless, it is still possible that assumptions A1, A2, or A3 may hold.

   Suppose your collaborators are interested in the ATT representing the effect of septic shock on Zubrod score one year after study entry, adjusting for confounding by baseline Zubrod score. Estimate the ATT using four different approaches: (1) risk difference with a DiD estimator and a linear model, (2) relative risk with a DiD estimator and a loglinear model, (3) odds ratio with a DiD estimator and a logistic model, and (4) using standardization with a saturated outcome model for all three estimates. Use the bootstrap to construct confidence intervals. Interpret your results. What do you tell your collaborators?

2. The hypothetical DiD Study sought to uncover the effect of treatment A on an biomarker measured before administration of treatment (Y_0) and one month following treatment administration (Y_1).

Suppose that $Y_0 = 1$ and $Y_1 = 1$ for the biomarker indicate a healthy condition. The data were generated with `ex2sim.r`:

```
ex2sim.r <- function() {
  set.seed(222)
  ss <- 1000
  Y_0 <- rbinom(n = ss, size = 1, prob = 0.5)
  probA <- 0.3 + 0.3 * Y_0
  A <- rbinom(n = ss, size = 1, prob = probA)
  probY_1 <- 0.2 + 0.2 * A + 0.2 * Y_0
  Y_1 <- rbinom(n = ss, size = 1, prob = probY_1)
  dat <- cbind(Y_0, A, Y_1)
  dat <- data.frame(dat)
  dat
}
```

Your job is to determine which analysis is valid: estimation of the ATT using the outcome-modeling approach to standardization, or estimation of the ATT using DiD estimation with a linear model. Use the `R` code to determine the true values of the estimands targeted by each approach. In addition, use `exsim.r` to simulate the data, and estimate the ATT as a risk difference using both methods. Interpret your results.

3. Suppose assumptions A1 and A2 both hold. What can you conclude about the relationships among the four components $a = E(Y_1(0)|A = 1)$, $b = E(Y_1(0)|A = 0)$, $c = E(Y_0|A = 1)$, and $d = E(Y_0|A = 0)$?

4. Suppose assumptions A2 and A3 both hold. What can you conclude about the relationships among the four components $a = E(Y_1(0)|A = 1)$, $b = E(Y_1(0)|A = 0)$, $c = E(Y_0|A = 1)$, and $d = E(Y_0|A = 0)$?

5. Suppose assumptions A1 and A3 both hold. What can you conclude about the relationships among the four components $a = E(Y_1(0)|A = 1)$, $b = E(Y_1(0)|A = 0)$, $c = E(Y_0|A = 1)$, and $d = E(Y_0|A = 0)$?

8

Adjusting for Confounding: Front-Door Method

8.1 Motivation

The front-door method of Pearl (1995) provides a way to adjust for confounding by unmeasured variables, represented by U, of the effect of A on Y in Figure 8.1. Like the backdoor method, the front-door method is quite clever, but to date, and unlike the backdoor method, it is not in popular use. Pearl and Mackenzie (2018) present just one published example of its use, by Glynn and Kashin (2018), who compare analyses of an observational study with randomized clinical trial (RCT) results in which the front-door method produces results that agree with those from the RCT but the backdoor method does not. However, in that example, as in so many others, it is difficult to justify the independence assumptions encoded in Figure 8.1; A indicates signing up for a job-training program offered by the Department of Labor, S indicates showing up for that program, and Y represents earnings over the subsequent 18 months. U indicates motivation, which is an unmeasured confounder. However, as there is likely an arrow from U to S, the front-door method would not be expected to work. Because it appears to work, either that arrow must be weak, or there must be differences between the RCT population and the observational study population that counterbalance it.

To motivate the front-door method, it is helpful to introduce the concept of a *surrogate marker*. In Figure 8.2, S is a surrogate marker for the effect of A on Y. Note that there may be other causes of Y that are independent of A and S, or causes of S that are independent of A; these need not be depicted in the DAG. If there are no such other causes, then S is a *perfect surrogate marker*, and testing the effect of A on Y is effectively identical to testing the effect of A on S. Surrogate markers are particularly useful when Y is a long-term clinical outcome and S occurs early on. For example, suppose we have a randomized clinical trial in a population of patients with high cholesterol, where A indicates initiation of treatment with a statin, S indicates a healthy lipid profile, and Y indicates subsequent myocardial infarction or stroke. In this example, S is not a perfect surrogate, because there are other causes of heart attacks and strokes besides unhealthy lipid levels, and possibly there are

DOI: 10.1201/9781003146674-8

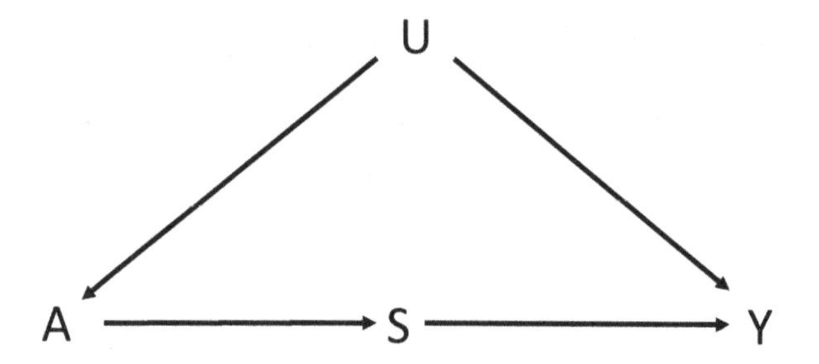

FIGURE 8.1: Front-door Causal DAG

also other causes of healthy lipid levels besides statins. However, it is plausible that statins prevent heart attacks and strokes solely by way of improving lipid levels. Initial trials could therefore focus solely on the surrogate marker, avoiding the lengthy trial duration a focus on clinical outcomes of real interest would require.

FIGURE 8.2: S is a Surrogate Marker

When S is a surrogate marker and the causal DAG of Figure 8.2 holds, we have that

$$E(Y|A) = \Sigma_s E(Y|S=s, A)P(S=s|A) = \Sigma_s E(Y|S=s)P(S=s|A),$$

because Y is independent of A given S. Additionally, as there are no confounders for the effect of A on Y, we have that

$$E(Y(a)) = E(Y|A=a) = \Sigma_s E(Y|S=s)P(S=s|A=a). \qquad (8.1)$$

Improved cholesterol can be thought of as a *specific* mechanism of the effect of statins; that is, it is the dominant effect of statins on the outcome. For treatments with more than one effect on the outcome, S could be a partial surrogate marker, as in Figure 8.3.

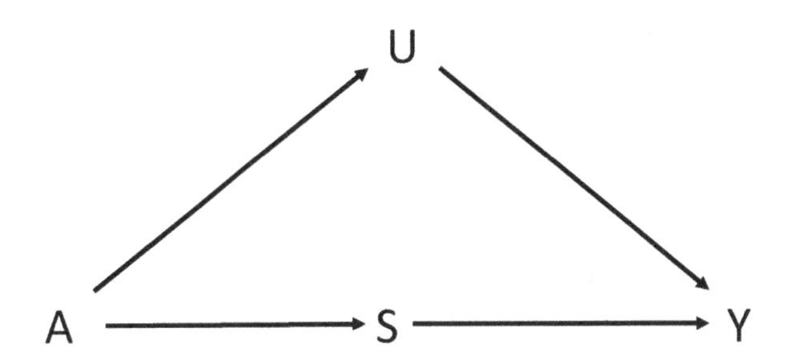

FIGURE 8.3: S is a Partial Surrogate Marker

The surrogate marker structure of Figure 8.2 plausibly arises in many RCTs of new therapeutic agents with specific mechanisms. Suppose such a trial confirmed a benefit of such a treatment in terms of improving the surrogate outcome, and that the treatment gained approval on this basis. A natural next step is to quantify the effect of the treatment on the clinical outcome of real interest. Although the RCT may be too small for this endeavor, a post-marketing observational study examining treatment use, surrogate outcome, and clinical outcome could conceivably be characterized by the front-door causal DAG in Figure 8.1. One needs to argue both that A does not affect Y except through S (i.e. that the mechanism of the therapeutic agent is specific), and that U does not affect S. In an observational study of statin use, restricted to a population with high cholesterol, U could represent family history of heart attack or stroke. This would plausibly affect statin use as well as subsequent heart attack or stroke. To use the front-door method, we would need to argue that it does not affect cholesterol reduction except via statin use. We would also need to rule out dietary restrictions as a successful alternative, which is perhaps somewhat plausible. Furthermore, we would need to consider other possible variables in U and also rule those out as direct causes of S.

The surrogate marker structure could instead pertain to unintended effects of a treatment. Second generation anti-psychotics (SGAs), which we code as A, are helpful treatments for mental illnesses including schizophrenia and bipolar disorder. However, they are well known to cause substantial weight gain S (Beasley et al. (1997)), with a risk of subsequent diabetes Y (Lambert et al.

(2006)). Nevertheless, there is evidence (Kohen (2004)) that schizophrenia and diabetes were associated long before the introduction of SGAs. Given electronic health record data on A, S, and Y from a mental health clinic, one might attempt to apply the front-door method to adjust for possible unmeasured confounders U, such as severity of schizophrenia. It is plausible that severity of schizophrenia would not cause substantial weight gain in the time-frame of the study, which was just one year in Lambert et al. (2006). Again, one would need to rule out other possible variables in U – affecting both SGA use and subsequent diabetes – as direct causes of substantial weight gain.

In much of science, we have much more information on the total effect of A on Y than we have on its mechanism. This could account, in large part, for the currently limited application of the front-door method. When we know the total effect but are interested in whether S serves as a mediator, i.e. in whether S is part of the mechanism, we would conduct a *mediation analysis.* Mediation analyses are the subject of Chapter 12.

8.2 Theory and Method

When the front-door causal DAG holds, the front-door theorm of Pearl (1995) states that we can modify equation (8.1) to

$$E(Y(a)) = \Sigma_s P(S=s|A=a)\Sigma_{a'} E(Y|S=s, A=a')P(A=a'), \qquad (8.2)$$

where we have replaced

$$E(Y|S=s)$$

with

$$\Sigma_{a'} E(Y|S=s, A=a')P(A=a'),$$

using backdoor standardization. The front-door method estimates $E(Y(a))$ via

$$\hat{E}(Y(a)) = \Sigma_s \hat{P}(S=s|A=a)\Sigma_{a'} \hat{E}(Y|S=s, A=a')\hat{P}(A=a').$$

We use a potential outcomes framework to prove the front-door theorem. We let $S(a)$ denote the potential outcome corresponding to S when A is set equal to a. We let $Y(a,s)$ denote the potential outcome corresponding to Y when A is set equal to a and then S is set equal to s. We let $Y(a)$ denote the potential outcome corresponding to Y when A is set equal to a. We then make the following six assumptions about these potential outcomes, where a and s represent any possible values of A and S.

1. We assume $Y(A) = Y$, $S(A) = S$, and $Y(A,S) = Y$. These are consistency assumptions.

2. $Y(a, S(a)) = Y(a)$, and $Y(a, S(a)) = Y(a, s)$ when $S(a) = s$. These are similar in spirit to consistency assumptions. The first states that had S been set equal to the value it would have taken were A set equal to a, we would have observed the potential outcome $Y(a)$.
3. $Y(a, s) = Y(\cdot, s)$, that is, it does not depend on a. We let $Y(\cdot, s)$ represent the potential outcome corresponding to Y when S is set equal to s. This assumption corresponds to no directed path from A to Y except through S. In particular, it rules out the partial surrogate DAG in Figure 8.3.
4. We assume that $S(a) \amalg A$, that is, there are no confounders for the effect of A on S.
5. We assume that $Y(\cdot, s) \amalg S|A$, that is, A is a sufficient confounder for the effect of S on Y.
6. We assume $Y(a, s) \amalg S(a)$. This is the key front-door assumption. It follows from the front-door causal DAG of Figure 8.1, where there is no arrow from U to S. It is perhaps easier to grasp from the version of the front-door causal DAG depicted in Figure 8.4, which includes the potential outcomes.

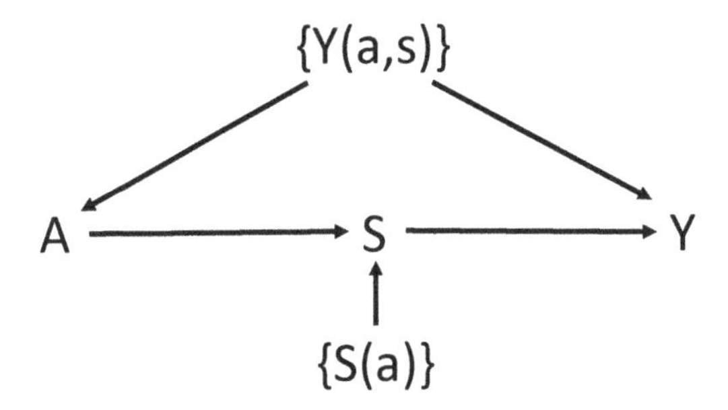

FIGURE 8.4: Front-Door Causal DAG Including Potential Outcomes

The proof of the front-door theorem at (8.2) is as follows.

$$E(Y(a)) = E(Y(a, S(a)))$$

by assumption 2.

$$E(Y(a, S(a))) = \Sigma_s E(Y(a, S(a))|S(a) = s)P(S(a) = s)$$

by the law of total expectation.

$$\Sigma_s E(Y(a, S(a))|S(a) = s)P(S(a) = s) = \Sigma_s E(Y(a, s)|S(a) = s)P(S(a) = s)$$

by assumption 2.

$$\Sigma_s E(Y(a,s)|S(a)=s)P(S(a)=s) = \Sigma_s E(Y(a,s))P(S(a)=s)$$

by the key front-door assumption 6.

$$\Sigma_s E(Y(a,s))P(S(a)=s) = \Sigma_s E(Y(\cdot,s))P(S(a)=s)$$

by assumption 3.

$$E(Y(\cdot,s)) = \Sigma_{a'} E(Y|S=s, A=a')P(A=a')$$

by assumption 5 and backdoor standardization.

$$P(S(a)=s) = P(S=s|A=a)$$

by assumption 4 and consistency. Then substitute the last two relations into the third to last relation to finish proving the front-door theorem.

8.3 Simulated Example

It is helpful to use a simulated binary dataset to better understand the theory and the method. One can think in terms of the SGA example, in which A indicates administration of SGA, S indicates substantial weight gain, and Y indicates diabetes at one year. We simulate according to the causal DAG in Figure 8.4. We let $P(Y(\cdot,0)=1)=0.05$ and $P(Y(\cdot,1)=1)=0.2$. We let $P(S(0)=1)=0.05$ and $P(S(1)=1)=0.9$. We let $P(A=1|Y(\cdot,0),Y(\cdot,1)) = (1-Y(\cdot,1))*0.1+Y(\cdot,1)*0.8$. Simulating these variables determines S and Y. Using R, the data are simulated with `sim1.r`:

```
sim1.r <- function ()
{
  set.seed(555)
  nsim = 10000
  # Generate the potential outcomes Y(.,0) and Y(.,1)
  Ydot0 <- rbinom(n = nsim, size = 1, prob = 0.05)
  Ydot1 <- rbinom(n = nsim, size = 1, prob = 0.2)
  # Let A depend on the Y(.,1)
  probA = (1 - Ydot1) * .1 + Ydot1 * .8
  A <- rbinom(n = nsim, size = 1, prob = probA)
  # Generate the potential outcomes S(0) and S(1)
  S0 <- rbinom(n = nsim, size = 1, prob = 0.05)
  S1 <- rbinom(n = nsim, size = 1, prob = 0.9)
  # S is a function of S(0), S(1), and A
  S <- (1 - A) * S0 + A * S1
  # Y is a function of Y(.,0), Y(.,1), and S
```

```
  Y <- (1 - S) * Ydot0 + S * Ydot1
  dat <- cbind(A, S, Y, Ydot0, Ydot1, S0, S1)
  data.frame(dat)
}
```

Using our theory, we can calculate

$$E(Y(a)) = E(Y(a, S(a))) = \Sigma_s E(Y(a, s))P(S(a) = s),$$

so that

$$E(Y(0)) = 0.05 * 0.95 + 0.2 * 0.05 = 0.0575$$

and

$$E(Y(1)) = 0.05 * 0.1 + 0.2 * 0.9 = 0.185.$$

We could instead try using equation (8.2) to calculate $E(Y(a))$, but this involves very many tedious applications of the multiplication rule and the law of total probability. Instead, we use `frontdoor.r` to estimate it, as follows.

```
frontdoor.r <- function ()
{
  # Estimate the components of E(Y(0))
  tmp00 <- (1 - mean(dat$S[dat$A == 0])) *
    (mean(dat$Y[(dat$S == 0) & (dat$A == 0)]) * (1 - mean(dat$A)) +
       mean(dat$Y[(dat$S == 0) & (dat$A == 1)]) * mean(dat$A))

  tmp01 <- (mean(dat$S[dat$A == 0])) *
    (mean(dat$Y[(dat$S == 1) & (dat$A == 0)]) * (1 - mean(dat$A)) +
       mean(dat$Y[(dat$S == 1) & (dat$A == 1)]) * mean(dat$A))

  EY0 <- tmp00 + tmp01

  #Estimate the components of E(Y(1))
  tmp10 <- (1 - mean(dat$S[dat$A == 1])) *
    (mean(dat$Y[dat$S == 0 & dat$A == 0]) * (1 - mean(dat$A)) +
       mean(dat$Y[dat$S == 0 & dat$A == 1]) * mean(dat$A))

  tmp11 <- mean(dat$S[dat$A == 1]) *
    (mean(dat$Y[dat$S == 1 & dat$A == 0]) * (1 - mean(dat$A)) +
       mean(dat$Y[dat$S == 1 & dat$A == 1]) * mean(dat$A))

  EY1 <- tmp10 + tmp11
  list(EY0 = EY0, EY1 = EY1)
}
> frontdoor.r()
$EY0
[1] 0.055847
$EY1
[1] 0.18537
```

We observe that the estimates are very close to the true values, as expected given the large sample size of 10,000. We can compare these estimates to the unadjusted estimates of $E(Y|A=0)$ and $E(Y|A=1)$, which are

```
> mean(dat$Y[dat$A == 0])
[1] 0.048892
> mean(dat$Y[dat$A == 1])
[1] 0.61367
```

We see that the estimate of $E(Y|A=1)$, in particular, is far away from $E(Y(1))$. There is substantial confounding, which the front-door method eliminated.

In summary, the front-door method represents a clever way to adjust for unmeasured confounding. However, the assumptions of the causal DAG in Figure 8.1 must hold. So far, the front-door method has not really found its way into applications. However, it is possible that with increasing awareness, it may prove useful for post-marketing follow-up of RCTs relying on surrogate marker outcomes, as in the statin example, or for safety analyses, as in the SGA example.

8.4 Exercises

1. Suppose the causal DAG of Figure 8.3 holds. Is equation (8.1) valid? Why or why not?
2. Simulate data according to `sim8ex2.r`. Let $Y(a)$ for $a = 0, 1$ be the potential outcome to $A = a$. Derive the true value of $E(Y(1)) - E(Y(0))$. Show empirically that applying equation (8.1) to estimate $E(Y(1)) - E(Y(0))$ leads to a biased estimator. Compare your results with estimation based on $E(Y(1)) - E(Y(0)) = E(Y|A=1) - E(Y|A=0)$. For both methods, use the bootstrap to construct confidence intervals.

```
sim8ex2.r <- function ()
{
  set.seed(8282)
  nsim <- 10000
  A <- rbinom(n = nsim, size = 1, prob = 0.5)
  probS <- .4 + .3 * A
  S <- rbinom(n = nsim, size = 1, prob = probS)
  probU <- .3 + .4 * A
  U <- rbinom(n = nsim, size = 1, prob = probU)
  probY <- .2 + .5 * U * S
  Y <- rbinom(n = nsim, size = 1, prob = probY)
  dat <- cbind(A, S, Y)
  data.frame(dat)
}
```

3. Simulate data according to `sim8ex3.r`. Let $Y(a)$ for $a = 0, 1$ be the potential outcome to $A = a$. Estimate $E(Y(1)) - E(Y(0))$ using the front-door approach and using standardization with an outcome model. Use the bootstrap to construct confidence intervals. Compare your results.

```
sim8ex3.r <- function ()
{
  set.seed(8383)
  nsim <- 10000
  X <- rbinom(n = nsim, size = 1, prob = 0.5)
  probA <- .3 + .4 * X
  A <- rbinom(n = nsim, size = 1, prob = probA)
  probS <- .4 + .3 * A
  S <- rbinom(n = nsim, size = 1, prob = probS)
  probY <- .2 + .5 * X * S
  Y <- rbinom(n = nsim, size = 1, prob = probY)
  dat <- cbind(X, A, S, Y)
  data.frame(dat)
}
```

4. Modify the key front-door assumption $Y(a, s) \amalg S(a)$ to show, mathematically, how to estimate $E(Y(0)|A = 1)$ assuming that the causal DAG of Figure 8.1 holds. How would you estimate the ATT?

5. Suppose the causal DAG of Figure 8.5 holds. Argue using potential outcomes that we can use the front-door method to estimate $E(Y(a))$. Use the front-door method to estimate the effect of AD_0 on A in the Double What-If? study.

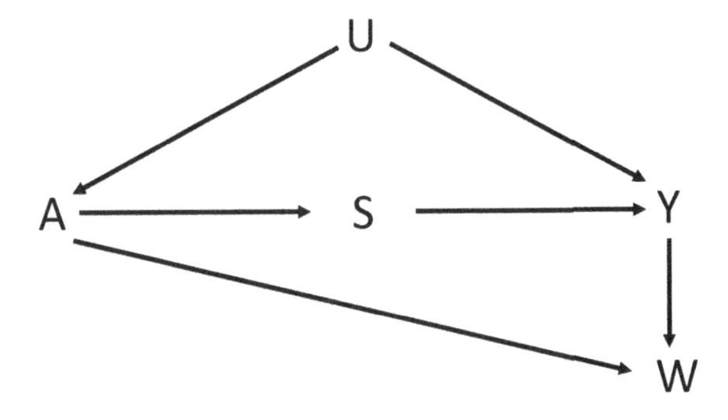

FIGURE 8.5: Causal DAG for Exercise 5

9

Adjusting for Confounding: Instrumental Variables

When the instrumental variables DAG of Figure 9.1 holds, we have yet another way to adjust for unmeasured confounding of the effect of A on Y. The variable T is as an *instrumental variable* if it (a) affects A, and (b) the only directed path from T to Y is through A, and (c) any backdoor paths from T to A or from T to Y are blocked by measured confounders in H. Assumption (b) is called the *exclusion assumption*, because it excludes the arrow directly from T to Y. When an instrumental variable exists, we can use it to adjust for unmeasured confounding using either *principal stratification* (Frangakis and Rubin (2002)) or *structural nested mean models* (Robins (1994) and Vansteelandt and Goetghebeur (2003)), under certain additional assumptions. In this chapter, we assume T is randomized so that there is no need for H, which we have therefore not included in the DAG. We also assume binary A and T. The methods in this chapter can be modified to include H or to include more general A and T; for example, see Brumback et al. (2014) and Helian et al. (2016).

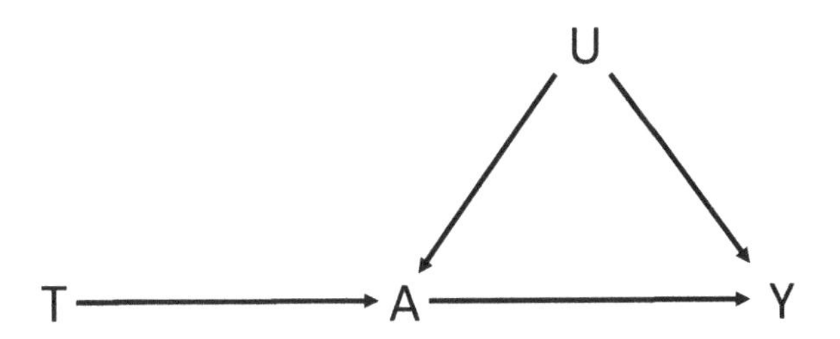

FIGURE 9.1: T is an Instrumental Variable for the Effect of A on Y

For example, in the What-If Study, we can use naltrexone T as an instrumental variable for the effect of reduced drinking A on unsuppressed viral load Y, provided that naltrexone does not cause unsuppressed viral load except via

DOI: 10.1201/9781003146674-9

reduced drinking. In the Double What-If Study, we can be certain that naltrexone T is an instrumental variable for the effect of reduced drinking A on unsuppressed viral load VL_1, because the data were generated such that T only causes VL_1 through A.

We enlist the following notation in this chapter. Let $Y(t, a)$ be the potential outcome for Y assuming we set $T = t$ and then $A = a$. We assume consistency. Due to the exclusion assumption, $Y(t, a) = Y(a)$. We let $A(t)$ be the potential outcome for A assuming we set $T = t$.

Suppose participants randomized to treatment with T either comply with the assigned treatment regime or not. Let the treatment actually taken be recorded in A. In the What-If and Double What-If Studies, we suppose that the entire purpose of treatment with naltrexone is to reduce drinking. Therefore we let reduced drinking equate to treatment actually taken. We note that A is a post-randomization event. When A does not equal T, two historical methods for assessing the effect of treatment on Y are called the *as-treated* analysis and the *per-protocol* analysis. In the as-treated analysis, we compare $E(Y|A = 1)$ with $E(Y|A = 0)$, and we lose the benefit of randomization. In the per-protocol analysis, we let $Z = 1$ if $A = T$ and use ordinary stratification on $Z = 1$ to compare $E(Y|T = 1, Z = 1)$ with $E(Y|T = 0, Z = 1)$. While this might look at first like a comparison of randomized groups, the event $Z = 1$ is a post-randomization event; therefore, Z, unlike a baseline covariate, cannot be expected to be balanced across the two treatment groups. For example, suppose sicker patients assigned to $T = 1$ comply, and have $A = 1$ and $Z = 1$, whereas some healthy patients avoid it due to side effects, and have $A = 0$ and $Z = 0$, while all patients assigned $T = 0$ comply with the assignment, and have $A = 0$ and $Z = 1$. In this case, while the as-treated and the per-protocol analyses are different, they both compare sicker patients on $T = 1$ with a blend of sicker and healthy patients on $T = 0$. Treatment will not appear as effective as it really is.

For these reasons, many studies rely on the *intent-to-treat effect* (ITT), $E(Y|T = 1) - E(Y|T = 0)$, which measures the effect of the intention to treat with $A = 1$ versus $A = 0$, rather than the effect of treatment actually received. Due to randomization and consistency, this equates to the causal effect

$$\text{ITT} = E(Y(1, A(1)) - E(Y(0, A(0)).$$

9.1 Complier Average Causal Effect and Principal Stratification

Principal stratification allows us to assess the causal effect of treatment actually received, under additional assumptions. Principal stratification classifies participants according to the potential occurence of a post-randomization event. It can be a useful alternative to ordinary stratification. With principal stratification, we define four principal strata of participants according to their potential outcomes $A(t)$. Following Angrist et al. (1996), if $A(0) = A(1) = 0$, then the participant is a *never taker*, and will not take the treatment regardless of randomized assignment. If $A(0) = A(1) = 1$, then the participant is an *always taker*, and will always take the treatment regardless of randomized assignment. If $A(0) = 0$ and $A(1) = 1$, then the participant is a *complier*, and will comply with whichever treatment was assigned. Finally, if $A(0) = 1$ and $A(1) = 0$, then the participant is a *defier*, and will refuse to comply with whichever treatment was assigned. Let $C = 1$ indicate a complier, i.e. that $A(t) = t$. Note that because (a) T is randomized, (b) C is a pre-randomization variable, and (c) $Y(t, a) = Y(a)$, it is reasonable to assume

$$Y(a) \amalg T|C, \tag{9.1}$$

that is, that T is independent of the potential outcomes $Y(a)$ within strata defined by C. If instead, $Y(t, a) \neq Y(a)$, then $Y(t, a)$ could lie on the path from T to A, in which case (9.1) would not hold. We also have that

$$T \amalg C$$

because C is a pre-randomization variable.

The *complier average causal effect* (CACE) is defined as the average effect of treatment in the compliers, that is, as

$$\begin{aligned} \text{CACE} &= E(Y(1)|C = 1) - E(Y(0)|C = 1) \\ &= E(Y|T = 1, C = 1) - E(Y|T = 0, C = 1), \end{aligned} \tag{9.2}$$

where the relation follows from consistency and assumption (9.1). The right-hand side of (9.2) shows the CACE as a stratified treatment effect, but $C = 1$ defines a principal stratum instead of an ordinary stratum. We cannot observe C for all participants. When $A(t) \neq t|T = t$ we know that the participant is a *non-complier*, but when $A(t) = t|T = t$ we do not know whether $A(1-t) = 1-t$ or not, and thus we cannot conclude that the participant is a complier. Furthermore, if we are willing to assume that there are no defiers, we can observe the proportion of always-takers as $E(A = 1|T = 0)$, and similarly we can observe the proportion of never-takers as $E(A = 0|T = 1)$, but we cannot observe the proportion of compliers $P(C = 1)$. Ordinary stratification assesses the treatment effect within a stratum defined by an observed variable, such as Z, whereas principal stratification assesses the treatment effect within a a stratum defined by an unobserved variable, such as C.

To estimate the CACE, Angrist et al. (1996) assume that there are no defiers, i.e. participants with $A(t) = 1 - t$. This is also referred to as the *monotonicity assumption*, because it is equivalent to $A(t) \geq A(t-1)$, e.g. that $A(t)$ is a monotonic function of t. Assuming exclusion and no defiers implies that

$$E(Y|T = 1, C = 0) = E(Y|T = 0, C = 0), \tag{9.3}$$

because no defiers means that $C = 0$ includes only never-takers and always takers, and exclusion ensures that randomization to T equal 1 or 0 cannot affect the outcome of the never-takers and the always-takers, because T does not affect A in those subgroups. Therefore, we have that

$$E(Y|T = 1) = E(Y|T = 1, C = 1)P(C = 1) + E(Y|T = 1, C = 0)(1 - P(C = 1)),$$

and

$$E(Y|T = 0) = E(Y|T = 0, C = 1)P(C = 1) + E(Y|T = 0, C = 0)(1 - P(C = 1)).$$

Therefore by (9.3),

$$E(Y|T = 1) - E(Y|T = 0) =$$
$$E(Y|T = 1, C = 1)P(C = 1) - E(Y|T = 0, C = 1)P(C = 1),$$

so that the CACE equals

$$E(Y|T = 1, C = 1) - E(Y|T = 0, C = 1) = \frac{E(Y|T = 1) - E(Y|T = 0)}{P(C = 1)}.$$

We note that no defiers implies

$$P(C = 1) + P(A = 0|T = 1) + P(A = 1|T = 0) = 1,$$

where the latter two quantities are the proportions of never takers and always takers, respectively. Thus,

$$\begin{aligned} P(C = 1) &= 1 - P(A = 0|T = 1) - P(A = 1|T = 0) \\ &= P(A = 1|T = 1) - P(A = 1|T = 0). \end{aligned}$$

Therefore, the CACE equals

$$\text{CACE} = \frac{E(Y|T = 1) - E(Y|T = 0)}{P(A = 1|T = 1) - P(A = 1|T = 0)}, \tag{9.4}$$

an expression phrased entirely in terms of observed variables. Therefore, we can estimate it via

$$\hat{\text{CACE}} = \frac{\hat{E}(Y|T = 1) - \hat{E}(Y|T = 0)}{\hat{P}(A = 1|T = 1) - \hat{P}(A = 1|T = 0)}. \tag{9.5}$$

The sampling distribution of $\hat{\text{CACE}}$ can be estimated using the bootstrap.

9.2 Average Effect of Treatment on the Treated and Structural Nested Mean Models

A drawback of the CACE is that it applies only to the compliers, a subgroup of the population that we cannot even identify. An alternative method targets the average effect of treatment on the treated (ATT), which we have already studied in Chapters 6 and 7:

$$ATT = E(Y(1) - Y(0)|A = 1) = E(Y - Y(0)|A = 1).$$

To estimate the ATT using the instrumental variable, T, we introduce structural nested mean models. We start with the linear structural nested mean model (SNMM):

$$E(Y - Y(0)|A, T) = A\beta. \tag{9.6}$$

We observe that $Y - Y(0)$ is assumed to be mean independent of T given A. This assumption for the ATT replaces the assumption of no defiers for the CACE. It implies that in the subset with $A = 1$, any effect modifiers of $Y - Y(0)$ are balanced across the $T = 0$ and $T = 1$ groups. This is a strong assumption. We also need the non-causal linear model

$$E(Y|A, T) = D\eta, \tag{9.7}$$

where D is a function of A and T. In this chapter, A and T are binary, and we can use the saturated model with $D = (1, A, T, A * T)$, so that this step does not add any additional assumptions. Combining (9.6) and (9.7), we have that

$$D\eta - A\beta = E(Y(0)|A, T),$$

and thus

$$E_{A|T}(D\eta - A\beta) = E(Y(0)|T) = E(Y(0)), \tag{9.8}$$

where the last equality is similar to (9.1), and results from T being randomized and from the exclusion assumption $Y(t, a) = Y(a)$, which entails that $Y(a)$ cannot lie on the path from T to A. Letting $E(Y(0)) = \alpha$, multiplying both sides by $(1, T)^T$, and then taking expectations of both sides of (9.8) results in the *instrumental variables estimating equation*

$$E\left[(1, T)^T(D\eta - A\beta - \alpha)\right] = 0. \tag{9.9}$$

To solve (9.9), we first estimate η by $\hat{\eta}$ using the methods of Chapter 2 with the linear model (9.7), and then we solve

$$\Sigma_{i=1}^{n}(1, T_i)^T(D_i\hat{\eta} - A_i\beta - \alpha) = 0 \tag{9.10}$$

for β and α using instrumental variables regression. The `ivreg` function in the `AER` library in `R` solves the estimating equation

$$\Sigma_{i=1}^{n}(1, T_i)^T(Y_i^* - A_i^*\beta - \alpha) = 0 \tag{9.11}$$

for β and α. Therefore, we can solve (9.10) by letting $Y^* = D\hat{\eta}$ and $A^* = A$. We provide examples using the software in the next section, but it is not hard to show that for binary T and A, the β which solves (9.11) is

$$\hat{ATT} = \hat{\beta} = \frac{\hat{E}(Y|T = 1) - \hat{E}(Y|T = 0)}{\hat{P}(A = 1|T = 1) - \hat{P}(A = 1|T = 0)}, \tag{9.12}$$

which happens to be identical to $\widehat{\text{CACE}}$ at (9.5). The interpretation, however, is quite different. Note that for the linear SNMM, $E(Y - Y(0)|A = 1, T) = \beta$ does not depend on T, and that is why β equals the ATT, $E(Y - Y(0)|A = 1)$.

The linear structural nested mean model can be generalized to

$$h(E(Y|A,T)) - A\beta = h(E(Y(0)|A,T)), \tag{9.13}$$

where $h(\cdot)$ is either $\log(\cdot)$ (for a loglinear structural nested mean model) or $\text{logit}(\cdot)$ (for a logistic structural nested mean model), and it is paired with the non-causal model

$$h(E(Y|A,T)) = D\eta, \tag{9.14}$$

where D is as before in (9.7). From (9.13) and (9.14) it follows that

$$h^{-1}(D\eta - A\beta) = E(Y(0)|A,T),$$

so that

$$E_{A|T}\left(h^{-1}(D\eta - A\beta)\right) = E(Y(0)|T) = E(Y(0)).$$

Letting $E(Y(0)) = \alpha$, multiplying both sides by $(1,T)^T$, and taking expectations of both sides yields

$$E\left[(1,T)^T\left(h^{-1}(D\eta - A\beta) - \alpha\right)\right] = 0. \tag{9.15}$$

To solve (9.15), we first solve the loglinear or logistic estimating equation associated with (9.14) to estimate η with $\hat{\eta}$, and then we solve (9.15) by iteratively linearizing the equation about current estimates β_t and α_t using a Taylor series approximation and solving an instrumental variables estimating equation of the form (9.11).

Specifically, when $h(\cdot) = \log(\cdot)$,

$$Y^* = \exp(D\hat{\eta} - A\beta_t)(1 + A\beta_t)$$

and

$$A^* = A\exp(D\hat{\eta} - A\beta_t).$$

When $h(\cdot) = \text{logit}(\cdot)$,

$$Y^* = \text{expit}(D\hat{\eta} - A\beta_t)\left(1 + A\beta_t(1 - \text{expit}(D\hat{\eta} - A\beta_t))\right)$$

and

$$A^* = A\text{expit}(D\hat{\eta} - A\beta_t)(1 - \text{expit}(D\hat{\eta} - A\beta_t)).$$

In the next section, we present R code that implements these estimators. We note that the general formulation applies to the linear structural nested mean model as well, with $h(\cdot)$ as the identity function.

Although the parameter β can be interpreted conditionally on $A = 1$ and T as a risk difference, log relative risk, or log odds ratio comparing the causal effect of $A = 1$ versus $A = 0$, we are interested in the ATT expressed as a risk difference, log relative risk, or log odds ratio, which is conditional on $A = 1$ but not on T. We have already shown that for the linear SNMM, β is the ATT expressed as a risk difference. For the loglinear SNMM, β is the log of the ATT expressed as a relative risk. To see this, let $h(\cdot)$ be the $\log(\cdot)$ function in equation (9.13) and let $A = 1$. We have that

$$\frac{E(Y|A=1,T)}{E(Y(0)|A=1,T)} = \exp(\beta).$$

Using the double expectation theorem,

$$\frac{E(Y|A=1)}{E(Y(0)|A=1)} = \frac{\Sigma_t E(Y|A=1,T=t)P(T=t|A=1)}{\Sigma_t E(Y(0)|A=1,T=t)P(T=t|A=1)},$$

which equals

$$\frac{\Sigma_t \exp(\beta)E(Y(0)|A=1,T=t)P(T=t|A=1)}{\Sigma_t E(Y(0)|A=1,T=t)P(T=t|A=1)} = \exp(\beta).$$

For all of the SNMMs, we can estimate $E(Y|A=1)$ simply as $\hat{E}(Y|A=1)$ since it is a function of observed data. For $E(Y(0)|A=1)$, we have

$$E(Y(0)|A=1) = E_{T|A=1}\left(h^{-1}(D\eta - A\beta)\right),$$

which we can estimate as the mean of $h^{-1}(D\hat{\eta} - A\hat{\beta})$ in the participants with $A = 1$. We also implement this with the `R` code of the next section.

The risk difference, log relative risk, and log odds ratio measures of the ATT are readily obtained from the estimates of $E(Y|A=1)$ and $E(Y(0)|A=1)$, as in Chapter 6. To estimate sampling distributions, we would typically turn to the bootstrap. However, the SNMM estimating equations are tricky. For the linear SNMM, the solution could be outside the range of $[-1, 1]$, which is non-sensical for risk differences, and for the loglinear and logistic SNMMs, sometimes the estimating equation has no solution. This occurs if the model does not fit the data very well. Even if the model fits the sampled data well, it may not fit data that are resampled using the bootstrap. Therefore, the bootstrap is problematic, particularly for the loglinear and logistic SNMMs. To circumvent this, we turn to the *jackknife* of Efron and Stein (1981), which is a simpler alternative to the bootstrap. The jackknife resamples the data $i = 1, \ldots, n$ times, where the ith resample includes the original sample except for participant i; then the resulting estimates are combined to obtain a good estimator of the standard error. By its very nature, the jackknife produces resampled datasets that are very close to the original data. Therefore, it would be highly unlikely to find a resampled dataset leading to failure of the loglinear and logistic SNMM estimating equations, provided that solutions exist for the original data. The `resample` package in `R` contains the `jackknife` function, which we use in the examples of the next section.

9.3 Examples

First, we consider the What-If? Study, in which we are interested in the effect of reduced drinking A on unsuppressed viral load Y. The `R` code below in `iv.r` and `bootiv.r` analyzes the data. A challenge with these data is that the proportion with $A = 1$ is almost the same in the $T = 0$ and $T = 1$ groups (62.35% versus 65%). When this occurs, T is called a *weak instrument* for the effect of A on Y. Furthermore, for some bootstrap samples, the proportions are identical and the denominator of the $\hat{CACE}$ and $\hat{ATT}$ equals zero, so that the estimator does not exist. For that reason, we use only 100 bootstrap samples instead of 1000, so that we could rerun the program until we obtained a group of 100 bootstrap samples with valid estimates. Typically, we would not conduct an analysis with a weak instrument, but we include it here for illustration of the difficulties.

```
> xtabs(~T+A,whatifdat)
   A
T    0  1
  0 32 53
  1 28 52
> prop.table(xtabs(~T+A,whatifdat),1)
   A
T         0       1
  0 0.37647 0.62353
  1 0.35000 0.65000
```

```
iv.r <- function(data = whatifdat,
                 ids = c(1:nrow(whatifdat)))
{
  dat <- data[ids, ]
  # Estimate the ITT
  ITT <- mean(dat$Y[dat$T == 1]) - mean(dat$Y[dat$T == 0])
  # Estimate the denominator of the IV estimator of the CACE and ATT
  denom <- mean(dat$A[dat$T == 1]) - mean(dat$A[dat$T == 0])
  IV <- ITT / denom
  c(ITT, IV)
}
bootiv.r <- function ()
{
  out <- boot(data = whatifdat,
              statistic = iv.r,
              R = 100)
  est <- summary(out)$original
  SE <- summary(out)$bootSE
  lci <- est - 1.96 * SE
  uci <- est + 1.96 * SE
  list(
    est = est,
```

```
    SE = SE,
    lci = lci,
    uci = uci
  )
}
> bootiv.r()
$est
[1] 0.0073529 0.2777778
$SE
[1]  0.071978 13.289620
$lci
[1]  -0.13372 -25.76988
$uci
[1]  0.14843 26.32543
```

We see that the ITT is positive, at 0.0074, but with a very wide confidence interval (−0.134, 0.148). The CACE and ATT are estimated at 0.278, which suggests that reduced drinking increases viral load, in contradiction to our results in Chapters 6 and 7. The weak instrument causes the discrepancy, and it also results in the very wide confidence interval of (−25.77, 26.32), which includes nonsensical risk differences.

The Double What-If Study provides a better basis for comparison of the CACE and ATT with the methods of Chapters 6 and 7. Furthermore, since we have the `R` code for the simulated data, we can compare to the true ATT risk difference, log relative risk, and log odds ratio, drawing on the calculations in Chapter 6. We can therefore determine whether our estimated ATT risk difference, log relative risk, and log odds ratio using the linear SNMM, loglinear SNMM, or logistic SNMM have confidence intervals that contain the truth. Due to the way we simulated the data, we do not know the true CACE. If we assume no defiers, we know that the true CACE equals the true ATT risk difference. However, we cannot tell from the simulation whether there are defiers or not.

We can determine from the `R` code for our simulation in `doublewhatifsim.r` that the linear SNMM holds but that the loglinear SNMM and the logistic SNMM do not hold. Let Y be `VL1`. We know from the simulation and the causal DAG that Y is independent of T given A and AD_0. We thus have that

$$E(Y(0)|A=1,T) = E_{AD_0|A=1,T}E(Y|A=0,AD_0)$$

and

$$E(Y|A=1,T) = E_{AD_0|A=1,T}E(Y|A=1,AD_0).$$

As in Chapter 6, we can determine from the `R` code that

$$E(Y|A,AD_0) = 0.855 - 0.36A - 0.4AD_0.$$

We also have that

$$\begin{aligned} E(AD_0|A=1,T) &= E_{U|A=1,T}E(AD_0|A=1,T,U) \\ &= E_{U|A=1,T}(0.2+0.6U) \\ &= 0.2+0.6E(U|A=1,T). \end{aligned}$$

We can use the multiplication rule to determine that

$$E(U|A=1,T) = P(U=1|A=1,T) = \frac{P(A=1,T|U=1)P(U=1)}{P(A=1,T)},$$

and the multiplication rule again plus that T is independent of U to determine that this equals

$$\frac{P(T)P(A=1|T,U=1)P(U=1)}{P(A=1,T)}.$$

Then by the law of total probabity, the denominator equals

$$P(A=1,T|U=0)P(U=0)+P(A=1,T|U=1)P(U=1).$$

Putting it altogether and using the probabilities from `doublewhatifsim.r`,

$$E(U|A=1,T) = \frac{0.5*0.5*(0.05+TU0.8)}{0.5(0.5*0.05+0.5(0.05+0.8T))},$$

so that

$$E(U|A=1,T) = \frac{0.0125+0.2T}{0.025+0.2T}.$$

Therefore,

$$E(AD_0|A=1,T=0) = 0.5$$

and

$$E(AD_0|A=1,T=1) = 0.7667.$$

Substituting, we find

$$\begin{aligned} E(Y(0)|A=1,T=0) &= 0.655 \\ E(Y(0)|A=1,T=1) &= 0.548 \\ E(Y|A=1,T=0) &= 0.295 \\ E(Y|A=1,T=1) &= 0.188 \end{aligned}$$

Finally, we can use these numbers to compute

$$\begin{aligned} E(Y-Y(0)|A=1,T=0) &= -0.36 \\ E(Y-Y(0)|A=1,T=1) &= -0.36, \end{aligned}$$

which means that the linear SNMM holds. However, as

$$\begin{aligned} \log(E(Y|A=1,T=0)) - \log(E(Y(0)|A=1,T=0)) &= -0.798 \\ \log(E(Y|A=1,T=1)) - \log(E(Y(0)|A=1,T=1)) &= -1.069, \end{aligned}$$

we observe that the loglinear SNMM does not hold. Similarly, as

$$\text{logit}(E(Y|A=1, T=0)) - \text{logit}(E(Y(0)|A=1, T=0)) = -1.512$$
$$\text{logit}(E(Y|A=1, T=1)) - \text{logit}(E(Y(0)|A=1, T=1)) = -1.655,$$

the logistic SNMM does not hold.

We compute the estimators for the linear, loglinear, and logistic SNMMs using the R functions `ividentity.r`, `ivlog.r`, `ivlogit.r`, shown below, and then we compute the jackknife confidence intervals using `jackiv.r`, also below.

```
ividentity.r <- function(data)
{
  dat <- data
  # Estimate D eta
  Deta <- predict(glm(VL1 ~ A * T, data = dat), type = "link")
  Ystar <- Deta
  Astar <- dat$A
  Z <- dat$T
  # Solve the IV estimating equation
  beta <- ivreg(formula = Ystar ~ Astar, instruments =  ~ Z)$coef[2]
  # Estimate E(Y(1)|A=1) and E(Y(0)|A=1)
  EY1 <- mean(Deta[dat$A == 1])
  EY0 <- mean((Deta - dat$A * beta)[dat$A == 1])
  # Estimate the effects
  RD <- EY1 - EY0
  logRR <- log(EY1 / EY0)
  logOR <- log(EY1 / (1 - EY1)) - log(EY0 / (1 - EY0))
  c(EY0, EY1, RD, logRR, logOR)
}
ivlog.r <- function(data)
{
  dat <- data
  niter = 10
  A <- dat$A
  Z <- dat$T
  # Estimate D eta
  Deta <- predict(glm(VL1 ~ A * T, family = poisson, data = dat),
       type = "link")
  # Initialize beta_t
  betat <- 0
  # Iteratively solve the IV estimating equation
  for (i in 1:niter)
  {
    #cat("i = ",i,"\n")
    # Find Ystar and Astar from Taylor series linearization
    Ystar <- exp(Deta - A * betat) * (1 + A * betat)
    Astar <- A * exp(Deta - A * betat)
    # Solve the IV estimating equation for the current iteration
    betat <- ivreg(formula = Ystar ~ Astar, instruments =  ~ Z)$coef[2]
    #cat("betat = ",betat,"\n")
  }
```

```
  beta <- betat
  # Compute estimates
  EY1 <- mean(exp(Deta)[A == 1])
  EY0 <- mean(exp(Deta - A * beta)[A == 1])
  RD <- EY1 - EY0
  logRR <- log(EY1 / EY0)
  logOR <- log(EY1 / (1 - EY1)) - log(EY0 / (1 - EY0))
  c(EY0, EY1, RD, logRR, logOR)
}
ivlogit.r <- function(data)
{
  dat <- data
  niter <- 10
  A <- dat$A
  Z <- dat$T
  # Estimate D eta
  Deta <- predict(glm(VL1 ~ A * T, family = binomial, data = dat),
                  type = "link")
  # Initialize beta_t
  betat <- 0
  # Iteratively solve the IV estimating equation
  for (i in 1:niter)
  {
    #cat("i = ",i,"\n")
    tmp <- exp(Deta - A * betat) / (1 + exp(Deta - A * betat))
    Ystar <- tmp * (1 + A * betat * (1 - tmp))
    Astar <- A * tmp * (1 - tmp)
    #return(ivreg(formula=Ystar~Astar,instruments=~Z))
    betat <- ivreg(formula = Ystar ~ Astar, instruments =  ~ Z)$coef[2]
    #cat("betat = ",betat,"\n")
  }
  beta <- betat
  # Compute estimates
  EY1 <- mean((exp(Deta) / (1 + exp(Deta)))[A == 1])
  EY0 <- mean((exp(Deta - A * beta) / (1 + exp(Deta - A * beta)))
                   [A == 1])
  RD <- EY1 - EY0
  logRR <- log(EY1 / EY0)
  logOR <- log(EY1 / (1 - EY1)) - log(EY0 / (1 - EY0))
  c(EY0, EY1, RD, logRR, logOR)
}
jackiv.r <- function ()
{
  out <- jackknife(data = doublewhatifdat, statistic = ividentity.r)
  est <- out$stats$Observed
  lci <- est - 1.96 * out$stats$SE
  uci <- est + 1.96 * out$stats$SE
  list(est = est, lci = lci, uci = uci)
}
```

The results are presented in Table 9.1. We see that all of the 95% confidence intervals for the SNMM methods contain the truth, despite the loglinear and logistic SNMMs not holding. The estimates from the loglinear and logistic SNMMs are indeed further from the truth than are those from the linear SNMM, as we would expect. However, it is reassuring that despite the model misspecification of the loglinear and logistic SNMM, the confidence intervals contain the true value. In practice, we have no way of knowing which, if any, of the SNMMs are correctly specified. We recommend trying all three, in hopes that the results qualitatively agree.

We conclude this chapter with one final cautionary example, which is hypothetical but nonetheless worth considering. Suppose $Y = 1$ indicates good health, $T = 1$ indicates a randomized new treatment with side effects, $T = 0$ indicates a placebo without side effects, and $A = 1$ indicates taking the new treatment. Suppose that individuals randomized to $T = 0$ cannot access the new treatment; this is not uncommon. Therefore all participants would adhere to the placebo $T = 0$ if assigned, so that for them, $A = 0$. Therefore, $P(A = 1|T = 0) = 0$. Suppose only 20% of those assigned the new treatment $T = 1$ adhered to it due to the side effects. Therefore, $P(A = 1|T = 1) = 0.2$ and $P(C = 1) = P(A = 1|T = 1) - P(A = 1|T = 0) = 0.2$. Note that there are no defiers.

Suppose there is a very strong placebo effect. Everyone assigned the treatment and taking the treatment experiences the placebo effect, as does everyone assigned the placebo and taking the placebo. To quantify it, suppose $E(Y|T = 1, C = 0) = 0.3$, whereas $E(Y|T = 0, C = 0) = E(Y|T = 0, C = 1) = E(Y|T = 1, C = 1) = 0.8$. Therefore, 80% of participants taking their assigned medication, whether the new treatment or placebo, experience good health, whereas only 30% of participants assigned to treatment but not taking it experience good health. Suppose further that $E(Y(0)|A = 1) = E(Y(0)|T = 1, C = 1) = 0.3$, so that only 30% of those taking the new treatment would have experienced good health if they had not taken it. With these numbers, we calculate that the true CACE is $E(Y|T = 1, C = 1) - E(Y|T = 0, C = 1) = 0$. To calculate the true ATT, we note that $E(Y|A = 1) = E(Y|T = 1, C = 1)$, because $A = 1$ exactly when $T = 1$ and $C = 1$. Therefore, the true ATT is $E(Y|A = 1) - E(Y(0)|A = 1) = 0.8 - 0.3 = 0.5$. Note that the linear SNMM holds automatically in this example, because $E(Y|A = 1, T = 1) = E(Y|A = 1)$ and $E(Y(0)|A = 1) = E(Y(0)|A = 1, T = 1)$, as A can equal one only when $T = 1$. Additionally, we automatically have $E(Y - Y(0)|A = 0, T) = 0$ due to consistency.

Because T is randomized, $P(C|T) = P(C)$, so that

$$
\begin{aligned}
E(Y|T = 1) &= E(Y|T = 1, C = 0)P(C = 0) + E(Y|T = 1, C = 1)P(C = 1) \\
&= 0.3 * 0.8 + 0.8 * 0.2 = 0.4.
\end{aligned}
$$

TABLE 9.1
Instrumental Variables Analysis of the Double What-If? Study

Measure	Truth	Method	Estimate (95% CI)
$E(Y(1) - Y(0) \mid C = 1)$	?	$\widehat{CACE}$	−0.409 (−0.564, −0.254)
$E(Y - Y(0) \mid A = 1)$	−0.360	Linear SNMM	−0.409 (−0.564, −0.254)
$E(Y - Y(0) \mid A = 1)$	−0.360	Loglinear SNMM	−0.483 (−0.746, −0.219)
$E(Y - Y(0) \mid A = 1)$	−0.360	Logistic SNMM	−0.413 (−0.560, −0.265)
$\log(E(Y \mid A = 1)) - \log(E(Y(0) \mid A = 1))$	−1.033	Linear SNMM	−1.020 (−1.340, −0.700)
$\log(E(Y \mid A = 1)) - \log(E(Y(0) \mid A = 1))$	−1.033	Loglinear SNMM	−1.129 (−1.564, −0.693)
$\log(E(Y \mid A = 1)) - \log(E(Y(0) \mid A = 1))$	−1.033	Logistic SNMM	−1.025 (−1.339, −0.712)
$\text{logit}(E(Y \mid A = 1)) - \text{logit}(E(Y(0) \mid A = 1))$	−1.630	Linear SNMM	−1.780 (−2.478, −1.081)
$\text{logit}(E(Y \mid A = 1)) - \text{logit}(E(Y(0) \mid A = 1))$	−1.630	Loglinear SNMM	−2.116 (−3.409, −0.822)
$\text{logit}(E(Y \mid A = 1)) - \text{logit}(E(Y(0) \mid A = 1))$	−1.630	Logistic SNMM	−1.794 (−2.464, −1.124)

Similarly, $E(Y|T=0)=0.8$. Therefore, the ITT is estimated at $0.4-0.8=-0.4$, and the CACE and ATT are estimated at

$$(E(Y|T=1)-E(Y|T=0))/P(A=1|T=1)=-0.4/0.2=-2,$$

an impossible value for a risk difference and thus a non-sensical result.

What has gone wrong? The placebo effect violates the exclusion assumption. Exclusion requires that $E(Y(t,a)) = E(Y(a))$, but in this example, $E(Y(1,0)) = 0.3$ whereas $E(Y(0,0)) = 0.8$. We see that the exclusion assumption excludes a placebo effect. In fact, in this example, the per-protocol analysis returns $E(Y|A=1,T=1)-E(Y|A=0,T=0)=0.8-0.8=0$, suggesting that treatment does nothing, which is the best answer.

This example highlights that assumptions cannot be made lightly, and that plausibility of exclusion must be carefully evaluated in any specific application. In more general settings in which the placebo group cannot access the treatment, one could also consider using measured variables to first predict compliance in the treatment group, and to then adjust the placebo group's outcomes using inverse-weighting, so that the weighted distribution approximates what would have been observed if the placebo group had consisted entirely of participants who would have adhered to the new treatment had it been assigned. The CACE could then be estimated as the difference between the mean outcome of compliers in the treatment group and the adjusted mean outcome in the placebo group. This method does not require the exclusion assumption to hold, and it reduces to the per-protocol analysis when there are no covariates. See Jo and Stuart (2009) for more details.

9.4 Exercises

1. Suppose we conduct a randomized trial to study the effect of treatment T on outcome Y, where A represents treatment actually taken, and that the causal graph of Figure 9.1 holds, which means that the exclusion assumption holds. Suppose further that those randomized to the placebo group $T=0$ have no access to the new treatment $T=1$; that is, A cannot equal 1 when $T=0$. Show that in this setting, the linear, loglinear, and logistic structural nested mean models impose no additional assumptions; that is, all three models automatically hold.

2. The dataset `vitaminadat` contains the data from a clinical trial of vitamin A supplementation to prevent death among children in Indonesia, analyzed by Sommer and Zeger (1991) and reproduced in Imbens and Rubin (1997). The variable T indicates randomization to vitamin A, the variable A indicates receipt of vitamin A, and

the variable Y indicates survival. Does anyone randomized to the control group receive vitamin A? Analyze the data using the CACE and also using the ATT via a linear SNMM. Interpret your results. What assumptions are required for validity of your interpretation?

3. Use the R code sim9ex3.r to generate the dataset sim9ex3dat.

```
sim9ex3.r <- function ()
{
  set.seed(9393)
  nsim = 50000
  U <- rbinom(n = nsim, size = 1, prob = 0.5)
  T <- rbinom(n = nsim, size = 1, prob = 0.5)
  probA <- U * T * 0.5 + 0.3
  A <- rbinom(n = nsim, size = 1, prob = probA)
  probS <- A * .8 + .1
  S <- rbinom(n = nsim, size = 1, prob = probS)
  meanY <- S * U
  Y <- rnorm(n = nsim, mean = meanY, sd = 1)
  dat <- cbind(U, T, A, S, Y)
  dat <- data.frame(dat)
  dat
}
```

Use the front-door approach, instrumental variables estimation with a linear SNMM, and standardization with an outcome model to estimate the ATT for the effect of A on Y. Use the bootstrap to estimate confidence intervals for each. What can you conclude about the validity of the linear SNMM in this example?

4. Use the R code sim9ex4.r to generate the dataset sim9ex4dat.

```
sim9ex4.r <- function ()
{
  beta <- 0.1
  set.seed(9494)
  nsim <- 5000
  Y0 <- rbinom(n = nsim, size = 1, prob = 0.5)
  T <- rbinom(n = nsim, size = 1, prob = 0.5)
  probA <- 0.2 * T + Y0 / 3 + .2
  A <- rbinom(n = nsim, size = 1, prob = probA)
  tmp <- (0.2 * T + 1 / 3 + .2) * 0.5 / (0.2 * T + 0.5 / 3 + .2)
  logitEY1 <- beta + log(tmp / (1 - tmp))
  EY1 <- exp(logitEY1) / (1 + exp(logitEY1))
  Y1 <- rbinom(n = nsim, size = 1, prob = EY1)
  Y <- Y0 * (1 - A) + Y1 * A
  dat <- cbind(Y0, Y1, T, A, Y)
  dat <- data.frame(dat)
  dat
}
```

Assume that the causal DAG of Figure 9.1 holds with $U = (Y(0), Y(1))$. Argue that the logistic SNMM is valid for data generated by `sim9ex4.r`. What does β represent? Estimate β and the ATT expressed as a risk difference, relative risk, and odds ratio, and compute jackknife confidence intervals for each. Does your jackknife confidence interval for β cover the true value?

5. Use the R code `sim9ex5.r` to generate the dataset `sim9ex5dat`.

```
sim9ex5.r <- function ()
{
  beta <- 4
  set.seed(9595)
  nsim <- 1000000
  Y0 <- rpois(n = nsim, lambda = 4)
  T <- rbinom(n = nsim, size = 1, prob = 0.5)
  probA <- 0.2 * T + Y0 / 40
  A <- rbinom(n = nsim, size = 1, prob = probA)
  logEYcA1T0 <- log(mean(Y0[(A == 1) & (T == 0)])) + beta
  logEYcA1T1 <- log(mean(Y0[(A == 1) & (T == 1)])) + beta
  nsim <- 5000
  Y0 <- rpois(n = nsim, lambda = 4)
  T <- rbinom(n = nsim, size = 1, prob = 0.5)
  probA <- 0.2 * T + Y0 / 40
  A <- rbinom(n = nsim, size = 1, prob = probA)
  Y1T0 <- rpois(n = nsim, lambda = exp(logEYcA1T0))
  Y1T1 <- rpois(n = nsim, lambda = exp(logEYcA1T1))
  Y <- Y0 * (1 - A) + Y1T0 * A * (1 - T) + Y1T1 * A * T
  dat <- cbind(T, A, Y)
  dat <- data.frame(dat)
  dat
}
```

Assume that the causal DAG of Figure 9.1 holds with $U = (Y(0), Y(1))$. Argue that the loglinear SNMM is valid for data generated by `sim9ex5.r`. What does β represent? Estimate β and the ATT expressed as a rate difference and rate ratio, and compute jackknife confidence intervals for each. Does your jackknife confidence interval for β cover the true value?

10

Adjusting for Confounding: Propensity-Score Methods

10.1 Theory

Chapter 6 defined the propensity score in the context of standardization using an exposure model. Since its introduction by Rosenbaum and Rubin (1983), the propensity score has served as the basis for several popular methods for confounding adjustment. Assuming

$$\{Y(t)\} \amalg T|H,$$

Rosenbaum and Rubin (1983) proved that

$$\{Y(t)\} \amalg T|e(H), \tag{10.1}$$

where $e(H)$ is the propensity score. This means that if H is a sufficient confounder, then so is $e(H)$. The theory holds exactly when $e(H)$ is a known function of H, but it is also approximately true when the parameters of $e(H)$ are estimated, for instance using a logistic model. This result conveniently implies that we need only work with the univariate confounder $e(H)$, rather than having to build models with the possibly complex confounder H. However, validity of the resulting methods requires that the model for $e(H)$ is correctly specified.

A cornerstone of propensity-score methods is *checking for overlap* of the distributions of the propensity score corresponding to the two treatment groups. To do this, we could compare the histograms for $e(H)$ across the two groups; however, it is somewhat easier to compare the probability densities for $e(H)$, where the probability density is essentially a smoothed version of the histogram. For illustration, we return to the General Social Survey data `gssrcc` to re-analyze the effect of completing more than high school on reporting a vote for Trump in 2016. We use the same confounders in H and the same logistic model for the propensity score as in Chapter 6.

Comparing the probability densities in Figure 10.1, we first observe that the density for `gthsedu=1` is higher for higher values of the propensity score than that for `gthsedu=0`. This just means that the participants with greater than

DOI: 10.1201/9781003146674-10

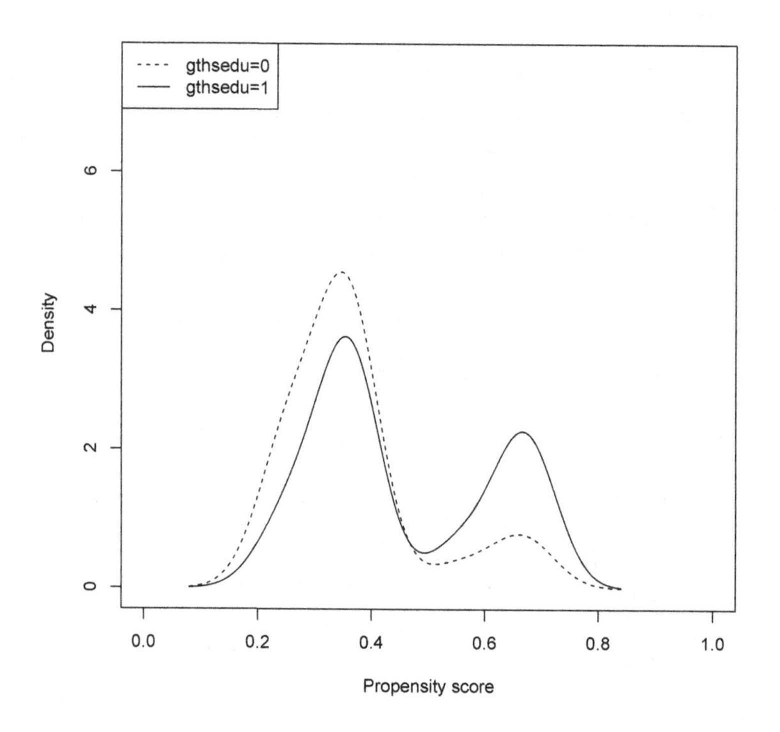

FIGURE 10.1: Checking for Overlap with the Trump Example

high school education tend to have a higher propensity for greater than high school education than those without greater than high school eduction, which is true by definition of the propensity score. Second, we observe that the densities overlap well; that is, given a participant with greater than high school education, we can find a participant without greater than high school education that has a similar propensity score, and vice versa. Suppose this were not the case, and there existed a group of participants with greater than high school education that had higher propensity scores than everyone without greater than high school education. Then these participants would have values of H very different from those of their counterparts without greater than high school education, and we would not know if a difference in outcomes were due to the difference in education or to the difference in confounders. This corresponds to a violation of the positivity assumption, because for that subgroup, $P(T = 0|H) = 0$. This phenomenon is illustrated in Figure 10.2, where we see a high density of treated individuals with propensity scores near one, unaccompanied by any untreated individuals.

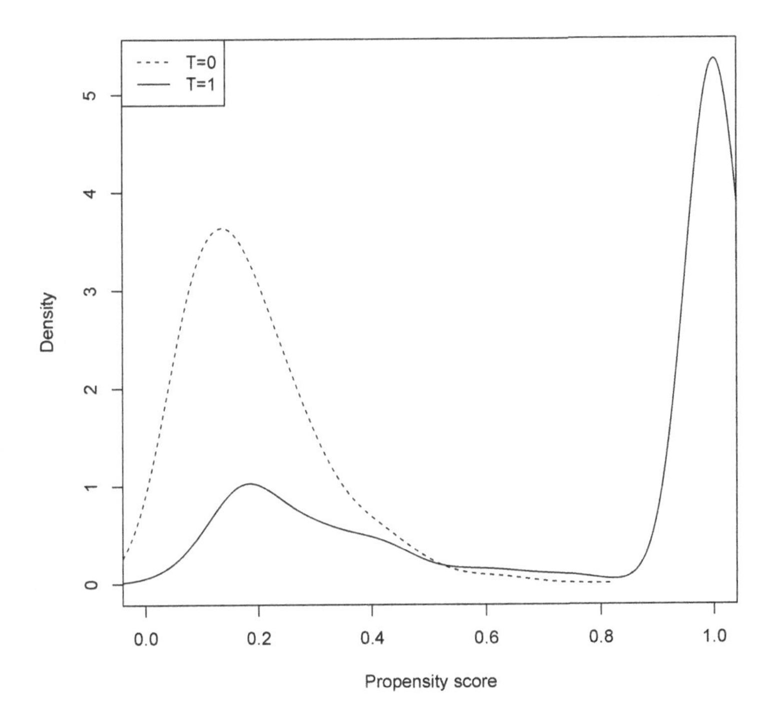

FIGURE 10.2: An Example of Overlap Failure

The data for Figure 10.2 were generated as follows:

```
nooverlap.r <- function()
{
  # Generate two confounders
  H1 <- rnorm(1000)
  H2 <- rbinom(n = 1000, size = 1, prob = .3)
  # Form a complex function of the confounders
  e <- exp(H1 + 3 * H2 - 1.5) / (1 + exp(H1 + 3 * H2 - 1.5))
  # Let T0 depend on the complex function
  T0 <- rbinom(n = 1000, size = 1, prob = e)
  # Let T1 equal H2
  T1 <- H2
  # Let the treatment be the maximum of T0 and T1
  T <- pmax(T0, T1)
  # Fit a propensity score model
  e <- fitted(glm(T ~ H1 + H2, family = binomial))
  # Find the range of the densities of the propensity score
  a <- range(density(e[T == 1], bw = .05)$y)
  b <- range(density(e[T == 0], bw = .05)$y)
  a <- range(a, b)
  # Set up the plot
  plot(c(0, 1),
       a,
       type = "n",
       xlab = "propensity score",
       ylab = "density")
  # Add the lines
  lines(density(e[T == 1], bw = .05), lty = 1)
  lines(density(e[T == 0], bw = .05), lty = 2)
  # Add a legend
  legend("topleft", c("T=0", "T=1"), lty = c(2, 1))
}
```

We see that T is heavily dependent on H_2, in that participants with $H_2 = 1$ are guaranteed to get the treatment.

When there is overlap failure, or equivalently violation of the positivity assumption, one solution is to exclude individuals in the treated group with $P(T = 0|H) = 0$ and individuals in the untreated group with $P(T = 1|H) = 0$ from the analysis. In so doing, the analysis would only apply to the subgroup of individuals with propensity scores in the region of overlap. A problem with this approach is that the subgroup may not be easy to characterize, and thus it might not be obvious to whom the treatment effect applies. However, it is likely better than any analyses that assume positivity when it is in fact untrue.

10.2 Using the Propensity Score in the Outcome Model

In Chapter 6, we defined the outcome model as a model for $E(Y|T,H)$. Due to (10.1), we could work with $e(H)$ instead of H, and hence with the outcome model $E(Y|T,e(H))$. We redo our outcome-model standardization of Chapter 6 with this reduced model, as follows:

```
estand.r <- function(data, ids)
{
  dat <- data[ids, ]
  # Fit the propensity score model
  emod <-
    glm(gthsedu ~ magthsedu + white + female + gt65,
        family = binomial,
        data = dat)
  e <- fitted(emod)
  # Fit the outcome model as a function of the propensity score
  lmod <- glm(trump ~ gthsedu + e, family = binomial, data = dat)
  dat0 <- dat1 <- dat
  dat0$gthsedu <- 0
  dat1$gthsedu <- 1
  dat0$e <- dat1$e <- e
  # Predict the potential outcomes for each participant
  EYhat0 <- predict(lmod, newdata = dat0, type = "response")
  EYhat1 <- predict(lmod, newdata = dat1, type = "response")
  # Estimate the average potential outcomes
  EY0 <- mean(EYhat0)
  EY1 <- mean(EYhat1)
  # Estimate the effects
  rd <- EY1 - EY0
  logrr <- log(EY1 / EY0)
  c(EY0, EY1, rd, logrr)
}
```

We also use the bootstrap as in Chapter 6. The results are presented in Table 10.1; they are almost identical to those of Table 6.8, computed using an ordinary outcome model.

Many researchers use the propensity score to estimate the conditional treatment effect instead of the standardized treatment effect. We can do this as follows:

```
Call:
glm(formula = trump ~ gthsedu + e, family = binomial, data = dat)

Deviance Residuals:
   Min       1Q  Median       3Q      Max
```

TABLE 10.1
Outcome-model Standardization Using the Propensity Score of the Effect of More than High School Education on Voting for Trump

Measure	Estimate	95% CI
$\hat{E}(Y(0))$	0.233	(0.210, 0.256)
$\hat{E}(Y(1))$	0.272	(0.243, 0.302)
$\hat{RD}$	0.040	(0.003, 0.077)
$\hat{RR}$	1.170	(1.015, 1.349)

```
-0.804  -0.787  -0.730  -0.718   1.722

Coefficients:
            Estimate Std. Error z value Pr(>|z|)
(Intercept)   -1.149      0.148   -7.77  7.6e-15
gthsedu        0.210      0.105    2.00    0.046
e             -0.110      0.360   -0.30    0.761
```

We see that the conditional odds ratio is exp(0.210) $=$ 1.2337, with a P-value of 0.046; a bootstrap P-value would be similar, but more accurate.

10.3 Stratification on the Propensity Score

The outcome model example above did not consider a possible interaction between T and $e(H)$. As such an interaction is always possible, Rosenbaum and Rubin (1983) recommend stratifying on the quintiles of the propensity score and computing the treatment effect within each quintile. We can also take an average of those treatment effects to obtain another standardized treatment effect, noting that an even 20% of the population falls into each quintile.

In our example, we instead stratify on quartiles $q_k(e)$, $k = 1, \ldots, 4$ of the propensity score. As H contains four binary variables, the propensity score does not take on very many unique values; in fact, there are only sixteen. We tried using quintiles, but while this works for point estimates, it fails for the bootstrap, because the five quintiles are not unique for some samples. We estimated the average potential outcomes in each quartile, $E(Y(0)|q_k(e))$ and $E(Y(1)|q_k(e))$, as well as the standardized risk difference, as follows:

```
equartiles.r <- function(data = gssrcc,
                         ids = c(1:nrow(gssrcc)))
{
  dat <- data[ids, ]
```

```
  # Fit the propensity score model using the prop.r function from
    Chapter 6
  eb <- prop.r(data, ids)$e
  # Find the quartiles of the propensity score
  quartiles <- quantile(eb, c(0, .25, .5, .75, 1))
  # Put participants into the quartiles
  equartiles <-
    cut(eb, breaks = quantile(eb, c(0, .25, .5, .75, 1)), include.
    lowest = T)
  # Estimate the average potential outcome within each quartile
  out <- glm(trump ~ gthsedu * equartiles - 1 - gthsedu, data = dat)
  # Extract the estimates
  EY0 <- out$coef[1:4]
  EY1 <- out$coef[1:4] + out$coef[5:8]
  # Estimate the risk difference
  RD <- mean(c(EY1 - EY0))
  RD
}
```

We also used the bootstrap with the standardized effect. The standardized risk difference was estimated at 0.030 (−0.017, 0.078). The point estimate is 25% lower than the estimate of the previous section, and it is no longer statistically significant. We graphed the estimated average potential outcomes in Figure 10.3.

Figure 10.3 explains why our standardized risk difference using stratification is lower than that for the outcome model ignoring interactions. In the third quartile, the risk difference is negative. Ignoring interactions smooths over this feature and results in a larger positive treatment effect.

10.4 Matching on the Propensity Score

Another way to adjust for confounding is to match each treated participant with one or more control participants having the same values for the confounders, and to similarly match each control particant with one or more treated participants. One can then calculate the risk difference or mean difference for each matched set and average across them to estimate the average treatment effect. Alternatively, one can estimate the ATT by restricting the matched sets to the matched treated participants. With more than one confounder, finding good matches is difficult if not impossible. The result at (10.1) implies that we only need to match on the propensity score. This approach to confounding adjustment has become very popular.

We illustrate in `R` with the `Match` function in the package `Matching`.

```
> summary(Match(Y=gssrcc$trump,Tr=gssrcc$gthsedu,X=e,
+ estimand="ATE",caliper=.25,replace=T,ties=F))
```

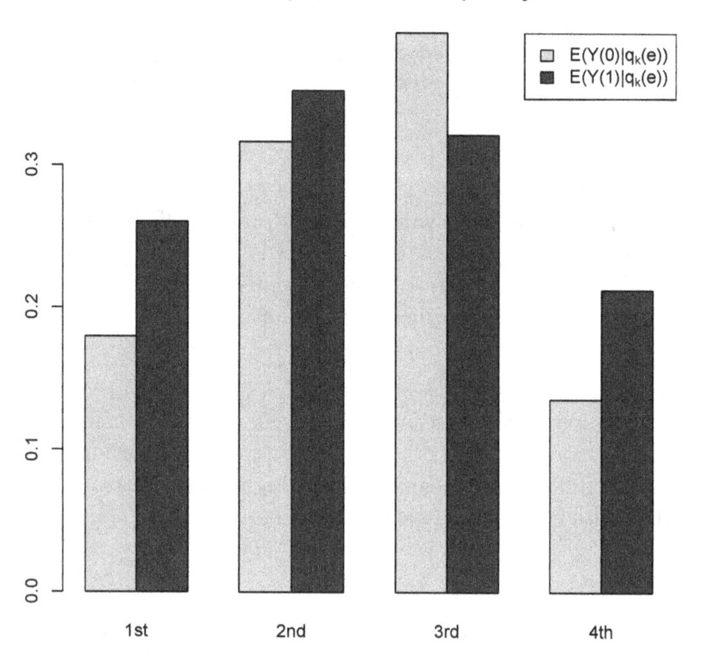

FIGURE 10.3: Average Potential Outcomes Within Strata of the Propensity Score

```
> Estimate...  0.0369
SE.........  0.012195
T-stat.....  3.0258
p.val......  0.00248

Original number of observations..............  2168
Original number of treated obs...............  874
Matched number of observations...............  2168
Matched number of observations  (unweighted).  2168

Caliper (SDs)........................................   0.25
Number of obs dropped by 'exact' or 'caliper'  0
> sqrt(var(e))
[1] 0.14401
```

We see that the estimated risk difference is 0.037, and that it is statistically significant. This is in between the standardized estimates using the ordinary and stratified approaches. We used a *caliper* of 0.25 standard deviations, or $0.25 * 0.144 = 0.036$ for matching the propensity scores. Thus, we have matched closely, and we see from the summary that every participant was matched. By changing `estimand="ATE"` to `estimand="ATT"` we obtain the ATT, also 0.037. The package also has a nice function for examining how well the matching has balanced the confounders across the treated and control groups:

```
> MatchBalance(trump~magthsedu+white+female+gt65,
+ data=gssrcc,match.out=match.out)

***** (V1) magthsedu *****
                       Before Matching          After Matching
mean treatment........     0.18182                 0.24493
mean control..........     0.26581                 0.24493
std mean diff.........     -21.756                       0

mean raw eQQ diff.....    0.083488                       0
med  raw eQQ diff.....           0                       0
max  raw eQQ diff.....           1                       0

mean eCDF diff........    0.041995                       0
med  eCDF diff........    0.041995                       0
max  eCDF diff........    0.083989                       0

var ratio (Tr/Co).....     0.76322                       1
T-test p-value........ 2.6725e-05                        1


***** (V2) white *****
                       Before Matching          After Matching
mean treatment........     0.95176                  0.7214
mean control..........     0.64518                  0.7214
```

```
std mean diff.........   142.95                           0

mean raw eQQ diff.....  0.30798                           0
med  raw eQQ diff.....        0                           0
max  raw eQQ diff.....        1                           0

mean eCDF diff........  0.15329                           0
med  eCDF diff........  0.15329                           0
max  eCDF diff........  0.30658                           0

var ratio (Tr/Co).....   0.2008                           1
T-test p-value........ < 2.22e-16                         1


***** (V3) female *****
                       Before Matching          After Matching
mean treatment........  0.46939                    0.55627
mean control..........  0.58502                    0.55627
std mean diff.........  -23.149                          0

mean raw eQQ diff.....  0.11503                          0
med  raw eQQ diff.....        0                          0
max  raw eQQ diff.....        1                          0

mean eCDF diff........ 0.057817                          0
med  eCDF diff........ 0.057817                          0
max  eCDF diff........  0.11563                          0

var ratio (Tr/Co).....   1.0272                          1
T-test p-value........ 3.4011e-06                        1


***** (V4) gt65 *****
                       Before Matching          After Matching
mean treatment........  0.31725                     0.2131
mean control..........  0.17864                    0.21218
std mean diff.........   29.756                    0.22523

mean raw eQQ diff.....  0.13915                 0.00092251
med  raw eQQ diff.....        0                          0
max  raw eQQ diff.....        1                          1

mean eCDF diff........ 0.069308                 0.00046125
med  eCDF diff........ 0.069308                 0.00046125
max  eCDF diff........  0.13862                 0.00092251

var ratio (Tr/Co).....   1.4781                     1.0032
T-test p-value........ 6.9199e-10                  0.15725
```

```
Before Matching Minimum p.value: < 2.22e-16
Variable Name(s): white  Number(s): 2

After Matching Minimum p.value: 0.15725
Variable Name(s): gt65  Number(s): 4
```

We observe from comparing `mean treatment` to `mean control` that all of the confounders are imbalanced before matching, but are nearly perfectly balanced after matching.

10.5 Exercises

1. Subset the `brfss` data introduced in the exercises for Chapter 3 and analyzed again in the exercises for Chapter 6 to include only those with `gt65` equal to zero. As we did in the exercises for Chapter 6, we will investigate the effect of `insured` on `flushot`, adjusting for the confounders `female`, `whitenh`, `blacknh`, `hisp`, `multinh`, `gthsedu`, and `rural`. First, estimate the propensity score and check for overlap of its distributions in the subgroups defined by `insured`.
2. Estimate the effect of `insured` on `flushot` conditional on the propensity score.
3. Use the outcome-modeling approach to standardization with the propensity score to estimate the effect of `insured` on `flushot`.
4. Apply stratification by quartiles of the propensity score to estimate the effect of `insured` on `flushot` within each quartile and also averaged across the quartiles. What happens when you try to use the bootstrap to estimate a confidence interval for the average? Which quartile boundaries are very close together? Merge those two quartiles and re-estimate the average and compute a bootstrap confidence interval for it.
5. Apply the `Match`, `summary.Match`, and `MatchBalance` functions to estimate the average treatment effect relating `insured` to `flushot` and to assess balance of the confounders.

11

Gaining Efficiency with Precision Variables

11.1 Theory

Although bias reduction of our estimators has been our focus thus far, we should nevertheless also reduce sampling variability when possible. In this chapter we introduce the concept of a *precision variable*, denoted by V in Figures 11.1 and 11.2, for reducing sampling varability of the estimated effect of a binary treatment T on an outcome Y.

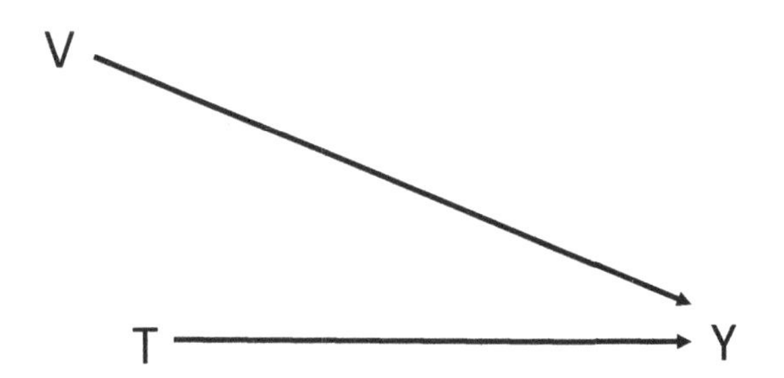

FIGURE 11.1: V is a Precision Variable for Estimating the Effect of T on Y

The key features of a precision variable are (1) that it is independent of the treatment and (2) that it is associated with the outcome.

A precision variable can reduce sampling variability in the estimated treatment effect by effectively subtracting variability in the outcome. For example, suppose the following models hold:

$$E(Y|T, V) = \beta_0 + \beta_1 T + \beta_2 V \tag{11.1}$$

so that

$$Y = \beta_0 + \beta_1 T + \beta_2 V + \epsilon. \tag{11.2}$$

DOI: 10.1201/9781003146674-11

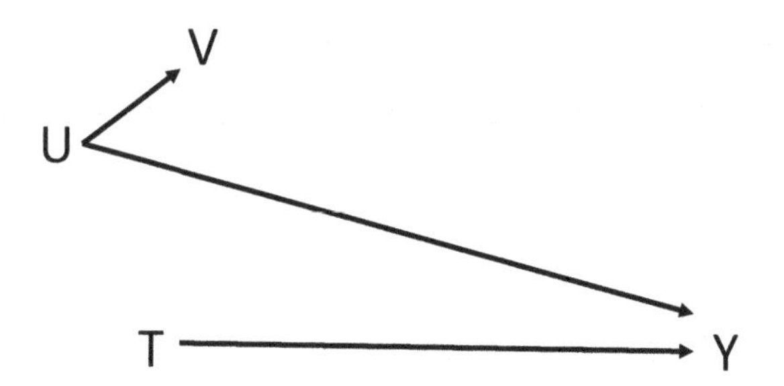

FIGURE 11.2: V is a Precision Variable for Estimating the Effect of T on Y

Suppose further that ϵ is independent of T and V and has variance equal to σ^2. By the double expectation theorem and the independence of V and T, the model at (11.1) also implies that the model

$$E(Y|T) = \alpha_0 + \beta_1 T \tag{11.3}$$

holds, where the coefficient of T is the same in the two models. This model in turn implies that

$$Y = \alpha_0 + \beta_1 T + \delta,$$

where

$$\delta = \beta_2(V - E(V)) + \epsilon.$$

Independence of ϵ and V implies that the variance of ϵ is less than the variance of δ, unless $\beta_2 = 0$. This is the key to using V to reduce sampling variability. Rather than using model (11.3) to estimate the treatment effect β_1, we use model (11.1) to produce a less variable estimator due to the smaller variance of ϵ, provided $\beta_2 \neq 0$. If $\beta_2 = 0$, the variances of the two estimators are effectively the same in large samples, but that of the estimator based on (11.1) is slightly more variable in small samples. However, this is a very small price to pay for the possibilty of a variance reduction due to an association of V with Y.

Due to independence of T and V, using model (11.1) to estimate the treatment effect produces an unbiased estimator of β_1 of model (11.3) even if model (11.1) does not hold. Whereas model (11.3) is saturated, model (11.1) makes parametric assumptions, which will not generally hold. However, using it for estimation still results in an estimator that is less variable than the one based on model (11.3) provided that V is associated with Y, and once again, we pay only a small price if V is not associated with Y.

It is somewhat difficult to see why using model (11.1) results in a less variable estimator than does model (11.3), even when (11.1) does not hold.

For those who are interested, we observe that the true model can be expressed as

$$E(Y|T,V) = \beta_0 + \beta_1 T + \beta_2 V + \gamma_3 f(V) + \gamma_4 g(V)T, \tag{11.4}$$

where $f(V)$ and $g(V)T$ have mean zero and are mean independent of T and V, and furthermore, $\beta_0, \beta_1, \beta_2$ are the same as in model (11.1). Let $\hat{\beta}_p$ be the estimator of β_1 using model (11.1) and $\hat{\beta}_u$ be the corresponding estimator using model (11.3). Then using the relation that

$$\text{var}(\hat{\theta}) = E\text{var}(\hat{\theta}|T,V) + \text{var}E(\hat{\theta}|T,V),$$

where $\hat{\theta}$ is any estimator and var$(\cdot)$ denotes the variance, we have that in large samples, $E(\text{var}(\hat{\beta}_p|T,V) = E(\text{var}(\hat{\beta}_u|T,V)$ but that

$$\text{var}E(\hat{\beta}_u|T,V) \geq \text{var}E(\hat{\beta}_p|T,V),$$

due to the latter variance being a function of $\gamma_3 f(V)+\gamma_4 g(V)T$ but the former being a function of $\beta_2 V + \gamma_3 f(V) + \gamma_4 g(V)T$, which is more variable due to the term $\beta_2 V$.

Therefore, when T is randomized, we recommend adopting the policy of choosing a precision variable and estimating the treatment effect using model (11.1), even when V is continuous and Y has a restricted range. Due to the likelihood that model (11.1) does not hold, sampling variability of the resulting estimator should be obtained using the bootstrap. One should select the precision variable prior to randomization, noting that model selection after the data are collected suffers from the problem of *multiple comparisons*. That is, suppose we had two possible precision variables V_1 and V_2. One might be tempted to fit model (11.1) twice, once with V_1 and once with V_2, and then to use the estimate of β_1 with a smaller standard error. This is similar to making two hypothesis tests about $\beta_1 = 0$, and rejecting the null if either test rejects it. The problem with this approach is that making more than one hypothesis test inflates the type I error. For example, if we make 20 or more independent hypothesis tests each with a type I error of 0.05, we can expect to reject at least one null hypothesis even when all the null hypotheses are true. This problem extends to model selection that considers interactions between V and T. That is why we generally recommend the simple policy above, unless prior studies strongly suggest an interaction or an effect of V^2, et cetera, in which case the one model considered most plausible from prior studies should be used.

11.2 Examples

For illustration, we consider two examples. First, we reanalyze data from a prospective, randomized, double-blind, placebo-controlled clinical trial designed to determine the effect of CYT107, the recombinant human IL-7 cytokine, on lymphocyte counts in sepsis (Francois et al. (2018)). We use absolute lymphocyte count (ALC) at day 29 as the outcome Y, and we let $T = 1$ for patients in either of two treatment groups, one with low frequency and one with high frequency administration of CYT107, for comparison with placebo. We let baseline ALC serve as the precision variable V. The data are stored in `il7dat`; we observe that it was a small trial, with only 10 participants in the placebo group and 27 in the two treatment groups combined. We analyzed the data with the functions `precision.r` and `bootprecision.r`; we used 100,000 replications of the bootstrap so that we could obtain very precise versions of the bootstrap standard errors.

```
precision.r <- function(data, ids)
{
  data = data[ids, ]
  # Estimate the effect without the precision variable
  TE1 <- summary(lm(Y ~ T, data = data))$coef[2]
  # Estimate the effect with the precicsion variable
  TE2 <- summary(lm(Y ~ T + V, data = data))$coef[2]
  c(TE1, TE2)
}
bootprecision.r <- function ()
{
  out <- boot(data = il7dat,
              statistic = precision.r,
              R = 100000)
  est <- summary(out)$original
  SE <- summary(out)$bootSE
  lci <- est - 1.96 * SE
  uci <- est + 1.96 * SE
  list(
    est = est,
    SE = SE,
    lci = lci,
    uci = uci
  )
}
```

The results are showin in Table 11.1. We observe that the standard error is very slightly larger when we include the precision variable using model (11.1) than when we do not include it using model (11.3). Therefore, in this study, rather surprisingly, baseline ALC is not a useful precision variable for the effect of CYT107 on 29 days ALC. However, we do not pay much of a price for using it for estimation. Both methods suggest that CYT107 is effective for increasing ALC in septic patients.

Looking at the output of fitting model (11.1), we see that the estimated coefficient of V is not statistically significant.

TABLE 11.1
Estimating the Effect of CYT107 on ALC with and without the Precision Variable Baseline ALC

Method	Estimate	SE	95% CI
Model (11.3)	1.349	0.476	(0.415, 2.28)
Model (11.1)	1.317	0.478	(0.380, 2.25)

```
> summary(lm(Y ~ T + V, data = il7dat))

Call:
lm(formula = Y ~ T + V, data = il7dat)

Residuals:
   Min     1Q Median     3Q    Max
-2.076 -0.772 -0.121  0.551  4.034

Coefficients:
            Estimate Std. Error t value Pr(>|t|)
(Intercept)    0.886      0.659    1.34    0.191
T              1.317      0.588    2.24    0.035
V              0.579      0.525    1.10    0.281

Residual standard error: 1.47 on 24 degrees of freedom
Multiple R-squared:  0.213,	Adjusted R-squared:  0.147
F-statistic: 3.25 on 2 and 24 DF,  p-value: 0.0565
```

For our second example, we analyze the effect of naltrexone T on unsuppressed viral load VL_1 using data from the Double What-If? Study. Even though we would recommend selecting a precision variable in advance, we present results of two possible choices, for teaching purposes. Without knowledge of the true causal DAG depicted in Figure 1.2, the obvious candidate would be VL_0, unsuppressed viral load at baseline. When a baseline version of the outcome is available, we would typically recommend choosing it. However in this unusual case, AD_0 is a better choice, because it directly causes VL_1, as in Figure 11.1, whereas VL_0 is only associated with VL_1 by way of AD_0, as in Figure 11.2. We simulated the Double What-If Study data so that the linear DiD estimator presented in Chapter 7 would be valid. This required us to use the causal DAG of Figure 1.2, rather than a more natural alternative. In real applications, we can never be sure of the causal DAG. Therefore, it is important to consider several plausible variants.

Using code similar to `precision.r` and `bootprecision.r`, we computed the estimates, standard errors, and confidence intervals shown in Table 11.2.

We observe that using VL_0 as the precision variable resulted in a negligible improvement, but using AD_0 reduced the standard error from 0.030 to

TABLE 11.2
Estimating the Effect of Naltrexone on Unsuppressed Viral Load Using the Double What-If? Data without a Precision Variable, with the Precision Variable VL_0, and with the Precision Variable AD_0

Precision Variable	Estimate	SE	95% CI
None	−0.147	0.031	(−0.208, −0.087)
VL_0	−0.146	0.030	(−0.205, −0.087)
AD_0	−0.151	0.026	(−0.203, −0.100)

0.026, and 13.3% reduction. Whereas in this study, incorporating AD_0 would not change the qualitative conclusion that naltrexone improved viral load, in other studies, a 13.3% reduction could make the difference between statistical significance or not. In still other studies, the reduction could be more pronounced.

11.3 Exercises

1. The `epil` dataset can be found in the `MASS` package of `R`:

```
> head(epil)
  y     trt base age V4 subject period    lbase     lage
1 5 placebo   11  31  0       1      1 -0.75635 0.114204
2 3 placebo   11  31  0       1      2 -0.75635 0.114204
3 3 placebo   11  31  0       1      3 -0.75635 0.114204
4 3 placebo   11  31  1       1      4 -0.75635 0.114204
5 3 placebo   11  30  0       2      1 -0.75635 0.081414
6 5 placebo   11  30  0       2      2 -0.75635 0.081414
```

The variable `y` is the outcome number of seizures recorded for four two-week periods (`period`) for epilepsy patients indexed by `subject` on progabide treatment (`trt`) or placebo. The variable `base` records the number of seizures in an eight-week baseline period. Construct a new dataset `epilave` by averaging `y` over the four periods, and explore whether using `base` as a precision variable reduces the sampling variability of your estimated treatment effect.

2. The `bacteria` dataset can be found in the `MASS` package of `R`:

```
> head(bacteria)
  y ap hilo week  ID     trt
1 y  p   hi    0 X01 placebo
2 y  p   hi    2 X01 placebo
3 y  p   hi    4 X01 placebo
```

```
4 y  p   hi   11 X01 placebo
5 y  a   hi    0 X02   drug+
6 y  a   hi    2 X02   drug+
```

The variable y indicates presence of the bacteria *H. influenzae* in children randomized to a drug or placebo (ap equal to a or p) at weeks 0, 2, 4, 6, and 11. Some weeks are missing. Week 0 gives the baseline presence or absence, which is prior to randomization. Construct a new dataset bact with the proportion of visits subsequent to week 0 with bacteria present as the outcome y and y0 equal to the baseline presence or absence. Explore whether using y0 as a precision variable reduces the sampling variability of your estimated treatment effect.

3. The toenail dataset comparing two randomized oral treatments for toenail infection can be found in the faraway package of R:

```
> head(toenail)
  ID outcome treatment    month visit
1  1       1         1  0.00000     1
2  1       1         1  0.85714     2
3  1       1         1  3.53571     3
4  1       0         1  4.53571     4
5  1       0         1  7.53571     5
6  1       0         1 10.03571     6
```

The variable outcome indicates the degree of separation of the toenail, with 0 indicating none or mild separation and 1 indicating moderate or severe. The variable treatment indicates the treatment received (A = 0 or B = 1). Visit 1 is baseline, prior to randomization. Construct a new dataset toe with the proportion of visits subsequent to visit 1 with moderate or severe separation as the outcome y and y0 equal to the baseline degree of separation. Explore whether using y0 as a precision variable reduces the sampling variability of your estimated treatment effect.

4. Generate a dataset using precisionsim.r and explore whether using V as a precision variable reduces the sampling variability of your estimated effect of T on Y. What is the true average treatment effect?

```
precisionsim.r <- function ()
  {
    nsim <- 90
    set.seed(123)
    # Generate a standard normal precision variable
    V <- rnorm(nsim)
    # Generate a Bernoulli treatment with probability 0.5
    T <- rbinom(n = nsim, size = 1, prob = 0.5)
    # Let the treatment effect depend on the precision variable
```

```
    EY <- .5 * T + T * V
    Y <- rnorm(n = nsim, mean = EY)
    dat <- cbind(V, T, Y)
    dat <- data.frame(dat)
    dat
}
```

5. Remove the `set.seed(123)` command in `precisionsim.r` and simulate 1000 datatsets. For each one, compute the length of the confidence intervals both including `V` as a precision variable and not including it, and also determine if the confidence interval covers the true average treatment effect. Does including `V` as a precision variable decrease the average length of the confidence intervals? Does it change the percentage of times the confidence interval covers the true average treatment effect? Construct confidence intervals for the coverage percentages to make sure they include 95%.

12

Mediation

Mediation analyses address questions about the mechanism M of the causal effect of A on Y. Does a new treatment A act by increasing absolute lymphocyte counts M which in turn improves Covid-19 outcomes Y? A difficulty with mediation analyses is that although we can randomize A, we typically cannot randomize M. Therefore, we must consider adjusting for confounding by H_2 of the effect of M on Y. Sometimes, we cannot randomize A either. In that case, we additionally must consider confounding by H_3 of the effect of A on M and confounding by H_1 of the effect of A on Y. Assumptions needed for confounding adjustment are depicted in the causal DAG of Figure 12.1. In the randomized trial, confounders H_1 and H_3 are absent, and we need only adjust for H_2. The next section presents the theory behind causal methods for mediation analyses assuming the causal DAG of Figure 12.1 holds and that the confounders H_1, H_2, and H_3 are observed. We reference VanderWeele and Vansteelandt (2009) and Vanderweele (2015) for these ideas and more about causal mediation.

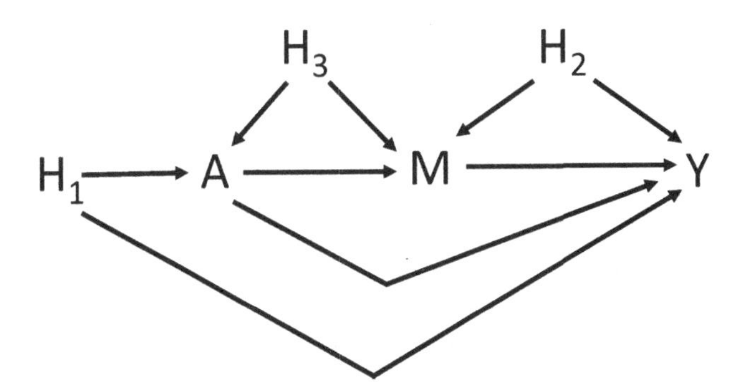

FIGURE 12.1: Causal DAG for Mediation by M of the Effect of A on Y

DOI: 10.1201/9781003146674-12

12.1 Theory

We introduce potential outcomes notation for both Y and M. We let $M(a)$ be the potential outcome of the mediator M to assigning treatment $A = a$. We let $Y(a, m)$ be the potential outcome for the outcome Y to assigning treatment $A = a$ and mediator $M = m$. We also consider potential outcomes of the form $Y(a, M(a^*))$, that is, we assign treatment $A = a$ and then we assign the mediator M to its potential outcome had we assigned treatment $A = a^*$. We make consistency assumptions for all potential outcomes. Analysis of mediation partitions the total effect into direct and indirect effects. The *total effect* (TE) compares the potential outcome for Y to assigning treatment $A = a$ with that of assigning treatment $A = a^*$:

$$TE(a, a^*) = Y(a, M(a)) - Y(a^*, M(a^*)).$$

The *controlled direct effect* (CDE) compares the potential outcome to assigning treatment $A = a$ with that of assigning treatment $A = a^*$ while controlling the mediator at $M = m$:

$$CDE(a, a^*, m) = Y(a, m) - Y(a^*, m).$$

The *controlled indirect effect* (CIE) is

$$CIE(a, a^*, m) = TE(a, a^*) - CDE(a, a^*, m).$$

An alternative partition does not control the mediator but rather lets it assume the value it would naturally take at a controlled level of A. The *natural direct effect* (NDE) compares the potential outcome to assigning treatment $A = a$ with that of assigning treatment $A = a^*$ while assigning $M = M(a^*)$:

$$NDE(a, a^*; a^*) = Y(a, M(a^*)) - Y(a^*, M(a^*)).$$

The *natural indirect effect* (NIE) compares the potential outcome to assigning $M = M(a)$ with that of assigning $M = M(a^*)$ while assigning $A = a$:

$$NIE(a, a^*; a) = Y(a, M(a)) - Y(a, M(a^*)).$$

Note that $TE = CDE + CIE$ and also that $TE = NDE + NIE$.

We cannot identify these effects on an individual level, but we can identify them on a population level in terms of their expected values. For estimation, we introduce the following four assumptions about the confounders:

1. There are no unmeasured treatment-outcome confounders given H (where $H = (H_1, H_2, H_3)$)
2. There are no unmeasured mediator-outcome confounders given (H, A)

3. There are no unmeasured treatment-mediator confounders given H
4. There is no mediator-outcome confounder affected by treatment (i.e. no arrow from A to H_2)

For the CDE and CIE, only assumptions (1) and (2) are needed. For the NDE and NIE, we need all four assumptions. Assumptions (1) and (3) are guaranteed when A is randomized.

To construct estimators of our causal quantities, we formalize these assumptions in terms of our potential outcomes:

1. $Y(a, m) \amalg A|H$
2. $Y(a, m) \amalg M|H, A$
3. $M(a) \amalg A|H$
4. $Y(a, m) \amalg M(a^*)|H$

Again, for the CDE and CIE, only (1) and (2) are needed, and once again, assumptions (1) and (3) are guaranteed when A is randomized.

Under assumption (1), the total effect conditional on H is given by

$$E(TE(a, a^*)|H = h) = E(Y|A = a, H = h) - E(Y|A = a^*, H = h).$$

We have already proved the two parts of this result in Chapter 3.

Under assumptions (1) and (2), the controlled direct effect conditional on H is given by:

$$\begin{aligned}&E(CDE(a, a^*; m)|H = h) =\\&E(Y|A = a, M = m, H = h) - E(Y|A = a^*, M = m, H = h).\end{aligned}$$

To prove this, recall that $CDE(a, a^*; m) = Y(a, m) - Y(a^*, m)$. By assumption (1), $E(Y(a, m)|H) = E(Y(a, m)|A, H)$. By assumption (2), $E(Y(a, m)|A, H) = E(Y(a, m)|A, M, H)$. By consistency, $E(Y(a, m)|A = a, M = m, H) = E(Y|A = a, M = m, H)$.

Under assumptions (1)-(4), the natural direct effect conditional on $H = h$ is

$$\begin{aligned}&E(NDE(a, a^*; a^*)|H = h) =\\&\Sigma_m \{E(Y|A = a, m, h) - E(Y|A = a^*, m, h)\} P(M = m|A = a^*, h). \quad (12.1)\end{aligned}$$

Under assumptions (1)–(4), the natural indirect effect conditional on $H = h$ is:

$$\begin{aligned}&E(NIE(a, a^*; a)|H = h) =\\&\Sigma_m E(Y|A = a, m, h)\{P(M = m|A = a, h) - P(M = m|A = a^*, h)\}. \quad (12.2)\end{aligned}$$

First, we prove (12.1).

1. Recall, $NDE(a, a^*; a^*) = Y(a, M(a^*)) - Y(a^*, M(a^*))$.
2. By the law of total expectation, $E(Y(a, M(a^*))|H) = \Sigma_m E(Y(a, M(a^*))|M(a^*) = m, H)P(M(a^*) = m|H)$.
3. By consistency, the right hand side equals $\Sigma_m E(Y(a, m)|M(a^*) = m, H)P(M(a^*) = m|H)$.
4. By assumptions (3) and (4), the right hand side equals $\Sigma_m E(Y(a, m)|H)P(M(a^*) = m|A = a^*, H)$.
5. By consistency and assumption (1), the right hand side equals $\Sigma_m E(Y(a, m)|A = a, H)P(M = m|A = a^*, H)$.
6. By assumption (2), the right hand side equals $\Sigma_m E(Y(a, m)|A = a, M = m, H)P(M = m|A = a^*, H)$.
7. By consistency, the right hand side equals $\Sigma_m E(Y|A = a, M = m, H)P(M = m|A = a^*, H)$.
8. Similarly, we can show $E(Y(a^*, M(a^*))|H) = \Sigma_m E(Y|A = a^*, M = m, H)P(M = m|A = a^*, H)$.

Second, we prove (12.2).

1. Recall, $NIE(a, a^*; a) = Y(a, M(a)) - Y(a, M(a^*))$.
2. By the previous proof we can show $E(Y(a, M(a))|H) = \Sigma_m E(Y|A = a, M = m, H)P(M = m|A = a, H)$.
3. By the previous proof we can also show $E(Y(a, M(a^*))|H) = \Sigma_m E(Y|A = a, M = m, H)P(M = m|A = a^*, H)$.
4. Thus

$$E(NIE(a, a^*; a)|H) =$$
$$\Sigma_m E(Y|A = a, M = m, H)\left\{P(M = m|A = a, H) - P(M = m|A = a^*, H)\right\}.$$

We can estimate all of these conditional effects by substituting parametric or nonparmetric estimators of the probabilities and expectations in the expressions involving the observed data. To estimate unconditional effects, we can average over the distribution of H in the population. For sampling variability, we can use the bootstrap.

We may often want to know how much of the total effect is mediated by M. Using the CDE and CIE partition, we can estimate the proportion of the effect that would be eliminated if we fixed the mediator to a specific level $M = m$:

$$PE = (TE - CDE(m))/TE.$$

Using the NDE and NIE partition, we can estimate the proportion mediated as

$$PM = NIE/TE.$$

In some applications, these measures can be less than zero or greater than one. In those cases, we do not use them, but rather restrict our description of the mediation to the CDE and CIE or the NDE and NIE.

In several applications, the dimension of H requires parametric methods. For example, we could use parametric models that accomodate treatment-mediator interaction:

$$E(Y|A = a, M = m, H = h) = \beta_0 + \beta_1 a + \beta_2 m + \beta_3 am + \beta_4^T h \tag{12.3}$$

$$E(M|A = a, H = h) = \alpha_0 + \alpha_1 a + \alpha_2^T h \tag{12.4}$$

Under assumptions (1) through (4), we can combine the estimates from the two models to obtain the following formulas for unconditional direct and indirect effects, comparing exposure levels a and a^*:

$$\begin{aligned} CDE(a, a^*; m) &= (\beta_1 + \beta_3 m)(a - a^*) \\ NDE(a, a^*; a^*) &= (\beta_1 + \beta_3(\alpha_0 + \alpha_1 a + \alpha_2^T E(H)))(a - a^*) \\ NIE(a, a^*; a) &= (\beta_2 \alpha_1 + \beta_3 \alpha_1 a)(a - a^*) \end{aligned}$$

12.2 Traditional Parametric Methods

Two traditional parametric methods assume models (12.3) and (12.4) but without an interaction in (12.3), that is, with $\beta_3 = 0$. In that case, we have

$$CDE(a, a^*; m) = \beta_1(a - a^*)$$

and

$$NIE(a, a^*; a) = \beta_2 \alpha_1 (a - a^*).$$

Letting $a = 1$ and $a^* = 0$, the NIE is estimated by $\hat{\beta}_2 \hat{\alpha}_1$. This is called the *product method* in the literature by Baron and Kenny (1986).

Alternatively, we can estimate the NIE by comparing the coefficient of a in

$$E(Y|A = a, H = h) = \beta_0 + \beta_1 a + \beta_2(\alpha_0 + \alpha_1 a + \alpha_2^T h) + \beta_4^T h,$$

which equals $\beta_1 + \beta_2 \alpha_1$, to the coefficient of a in

$$E(Y|A = a, M = m, H = h) = \beta_0 + \beta_1 a + \beta_2 m + \beta_4^T h,$$

which equals β_1. The idea is to see if the coefficient $\beta_1 + \beta_2 \alpha_1$ is significantly reduced by adding M to the regression. We see it is reduced by the difference of those two coefficients, which is $\beta_2 \alpha_1$, which equals the NIE of the product method. This method is known as the *difference method* in the literature; for example, see Jiang and VanderWeele (2015). We estimate sampling variability using the bootstrap.

To illustrate these two methods, we use the General Social Survey data to investigate whether the effect of maternal education of greater than high school on a reported vote for Trump is mediated by conservative political beliefs of the respondent. For simplicity, we omit H from this analysis. We used the dataset `gssmed`, which is the same as `gssrcc` but with conservative beliefs added and restricted to the 2084 complete cases. The results of fitting the three models are as follows.

```
> summary(glm(trump ~ magthsedu, data = gssmed))
Call:
glm(formula = trump ~ magthsedu, data = gssmed)
Coefficients:
            Estimate Std. Error t value Pr(>|t|)
(Intercept)  0.2763     0.0110   25.18  < 2e-16
magthsedu   -0.0912     0.0219   -4.17  3.2e-05
> summary(glm(trump ~ conservative + magthsedu, data = gssmed))
Call:
glm(formula = trump ~ conservative + magthsedu, data = gssmed)
Coefficients:
             Estimate Std. Error t value Pr(>|t|)
(Intercept)    0.1436     0.0118   12.22  < 2e-16
conservative   0.3906     0.0184   21.23  < 2e-16
magthsedu     -0.0688     0.0199   -3.46  0.00055
> summary(glm(conservative ~ magthsedu, data = gssmed))
Call:
glm(formula = conservative ~ magthsedu, data = gssmed)
Coefficients:
            Estimate Std. Error t value Pr(>|t|)
(Intercept)  0.3397     0.0119   28.67   <2e-16
magthsedu   -0.0573     0.0236   -2.42    0.015
```

We see that the product method estimates the NIE as $0.391*(-0.057) = -0.022$, whereas the difference method estimates it as $-0.091 - (-0.069) = -0.022$; the two methods yield the same result. We assessed statistical significance using the bootstrap with the difference method, which produced a confidence interval of $(-0.040, -0.005)$. Thus, the mediation is statistically significant.

12.3 More Examples

We present two more examples, letting $a = 1$ and $a^* = 0$ in the estimation of the unconditional TE, CDE, CIE, NDE, and NIE. First, we investigate whether the effect of naltrexone on unsuppressed viral load is mediated by reduced drinking, adjusting for confounding by baseline antiretroviral adherence, using data from the Double What-If? Study. We reassign variables so that A is naltrexone, H is baseline adherence, M is reduced drinking, and Y is unsuppressed viral load. For teaching purposes, we compute the causal estimates two different ways. First, we use `mediation.r`:

```
mediation.r <-
   function(dat = meddoublewhatifdat, ids = c(1:nrow(meddoublewhatifdat)))
   {
     dat <- dat[ids, ]
     # Estimate P(H=0) and P(H=1)
     PH0 <- 1 - mean(dat$H)
     PH1 <- mean(dat$H)
     # Estimate the first term of equation (1) as a function of H
     EY1M0H0 <- (mean(dat$Y[(dat$A == 1) & (dat$M == 0) & (dat$H == 0)]) *
                   (1 - mean(dat$M[(dat$A == 0) & (dat$H == 0)]))
                 + mean(dat$Y[(dat$A == 1) & (dat$M == 1) & (dat$H == 0)]) *
                   (mean(dat$M[(dat$A == 0) & (dat$H == 0)])))
     EY1M0H1 <- (mean(dat$Y[(dat$A == 1) & (dat$M == 0) & (dat$H == 1)]) *
                   (1 - mean(dat$M[(dat$A == 0) & (dat$H == 1)]))
                 + mean(dat$Y[(dat$A == 1) & (dat$M == 1) & (dat$H == 1)]) *
                   (mean(dat$M[(dat$A == 0) & (dat$H == 1)])))
     # Standardize with respect to marginal distribution of H
     EY1M0 <- EY1M0H0 * PH0 + EY1M0H1 * PH1
     # Estimate the second term of equation (1) as a function of H
     EY0M0H0 <- (mean(dat$Y[(dat$A == 0) & (dat$M == 0) & (dat$H == 0)]) *
                   (1 - mean(dat$M[(dat$A == 0) & (dat$H == 0)]))
                 + mean(dat$Y[(dat$A == 0) & (dat$M == 1) & (dat$H == 0)]) *
                   (mean(dat$M[(dat$A == 0) & (dat$H == 0)])))
     EY0M0H1 <- (mean(dat$Y[(dat$A == 0) & (dat$M == 0) & (dat$H == 1)]) *
                   (1 - mean(dat$M[(dat$A == 0) & (dat$H == 1)]))
                 + mean(dat$Y[(dat$A == 0) & (dat$M == 1) & (dat$H == 1)]) *
                   (mean(dat$M[(dat$A == 0) & (dat$H == 1)])))
     # Standardize with respect to marginal distribution of H
     EY0M0 <- EY0M0H0 * PH0 + EY0M0H1 * PH1
     # Estimate the first tirm of equation (2) as a function of H
     EY1M1H0 <- (mean(dat$Y[(dat$A == 1) & (dat$M == 0) & (dat$H == 0)]) *
                   (1 - mean(dat$M[(dat$A == 1) & (dat$H == 0)]))
                 + mean(dat$Y[(dat$A == 1) & (dat$M == 1) & (dat$H == 0)]) *
                   (mean(dat$M[(dat$A == 1) & (dat$H == 0)])))
     EY1M1H1 <- (mean(dat$Y[(dat$A == 1) & (dat$M == 0) & (dat$H == 1)]) *
                   (1 - mean(dat$M[(dat$A == 1) & (dat$H == 1)]))
                 + mean(dat$Y[(dat$A == 1) & (dat$M == 1) & (dat$H == 1)]) *
                   (mean(dat$M[(dat$A == 1) & (dat$H == 1)])))
     # Standardize with respect to marginal distribution of H
     EY1M1 <- EY1M1H0 * PH0 + EY1M1H1 * PH1
     # Estimate the CDE setting M=1 as a function of H
     CDE1H0 <- (mean(dat$Y[(dat$A == 1) & (dat$M == 1) & (dat$H == 0)])
               - mean(dat$Y[(dat$A == 0) & (dat$M == 1) & (dat$H == 0)]))
     CDE1H1 <- (mean(dat$Y[(dat$A == 1) & (dat$M == 1) & (dat$H == 1)])
               - mean(dat$Y[(dat$A == 0) & (dat$M == 1) & (dat$H == 1)]))
     # Standardize with respect to the marginal distribution of H
     CDE1 <- CDE1H0 * PH0 + CDE1H1 * PH1
     # Estimate the CDE setting M=0 as a function of H
     CDE0H0 <- (mean(dat$Y[(dat$A == 1) & (dat$M == 0) & (dat$H == 0)])
               - mean(dat$Y[(dat$A == 0) & (dat$M == 0) & (dat$H == 0)]))
     CDE0H1 <- (mean(dat$Y[(dat$A == 1) & (dat$M == 0) & (dat$H == 1)])
               - mean(dat$Y[(dat$A == 0) & (dat$M == 0) & (dat$H == 1)]))
```

```
    # Standardize with respect to the marginal distribution of H
    CDE0 <- CDE0H0 * PH0 + CDE0H1 * PH1
    # Estimate the natural direct effect
    NDE <- EY1M0 - EY0M0
    # Estimate the natural indirect effect
    NIE <- EY1M1 - EY1M0
    # Estimate the total effect
    TE <- NDE + NIE
    # Estimate the proportion mediated
    PM <- (NIE / TE)
    # Estimate the proportion eliminated setting M=1
    PE1 <- ((TE - CDE1) / TE)
    # Estimate the proportion eliminated setting M=0
    PE0 <- ((TE - CDE0) / TE)
    c(TE, CDE0, CDE1, NDE, NIE, PE0, PE1, PM)
  }
```

Second, we use `nonparamediation.r`:

```
nonparamediation.r <-
  function(dat = meddoublewhatifdat, ids = c(1:nrow(meddoublewhatifdat)))
  {
    dat <- dat[ids, ]
    # Fit a saturated outcome model including the mediator
    out <- glm(Y ~ A * M * H, family = binomial, data = dat)
    # Fit a saturated model for the mediator
    med <- glm(M ~ A * H, family = binomial, data = dat)
    dat10 <- dat00 <- dat01 <- dat11 <- dat
    dat10$A <- 1
    dat10$M <- 0
    dat00$A <- 0
    dat00$M <- 0
    dat01$A <- 0
    dat01$M <- 1
    dat11$A <- 1
    dat11$M <- 1
    # Estimate the first term of equation (1) as a function of H
    EY1M0H <- (
      predict(out, newdata = dat10, type = "response") *
        (1 - predict(
          med, newdata = dat00, type = "response"
        ))
      + predict(out, newdata = dat11, type = "response") *
        predict(med, newdata = dat01, type = "response")
    )
    # Standardize with respect to the marginal distribution of H
    EY1M0 <- mean(EY1M0H)
    # Estimate the second term of equation (1) as a function of H
    EY0M0H <- (
      predict(out, newdata = dat00, type = "response") *
        (1 - predict(
          med, newdata = dat00, type = "response"
        ))
```

```
    + predict(out, newdata = dat01, type = "response") *
      predict(med, newdata = dat01, type = "response")
  )
  # Standardize with respect to the marginal distribution of H
  EY0M0 <- mean(EY0M0H)
  # Estimate the first term of equation (2) as a function of H
  EY1M1H <- (
    predict(out, newdata = dat10, type = "response") *
      (1 - predict(
        med, newdata = dat10, type = "response"
      ))
    + predict(out, newdata = dat11, type = "response") *
      predict(med, newdata = dat11, type = "response")
  )
  # Standardize with respect to the marginal distribution of H
  EY1M1 <- mean(EY1M1H)
  # Estimate the controlled direct effects
  CDE1H <- (
    predict(out, newdata = dat11, type = "response") -
      predict(out, newdata = dat01, type = "response")
  )
  CDE1 <- mean(CDE1H)
  CDE0H <- (
    predict(out, newdata = dat10, type = "response") -
      predict(out, newdata = dat00, type = "response")
  )
  CDE0 <- mean(CDE0H)
  # Estimate the natural direct and indirect effects
  NDE <- EY1M0 - EY0M0
  NIE <- EY1M1 - EY1M0
  # Estimate the total effect
  TE <- NDE + NIE
  # Estimate the proportion mediated
  PM <- (NIE / TE)
  # Estimate the proportions eliminated
  PE1 <- ((TE - CDE1) / TE)
  PE0 <- ((TE - CDE0) / TE)
  c(TE, CDE0, CDE1, NDE, NIE, PE0, PE1, PM)
}
```

These two functions both implement a nonparametric mediation analysis and return exactly the same answers, presented in Table 12.1. The second function is easily modified to incorporate a higher-dimensional H and also parametric modeling assumptions. Table 12.2 presents results of a parametric mediation analysis, letting

```
out <- glm(Y ~ A + M + H, family = binomial, data = dat)
```

and

```
med <- glm(M ~ A + H, family = binomial, data = dat)
```

in the second function.

TABLE 12.1
Nonparametric Mediation Analysis of the Double What-If? Study

Measure	Estimate	95% CI
TE	−0.151	(−0.202, −0.101)
CDE(0)	−0.042	(−0.111, 0.026)
CDE(1)	0.047	(−0.123, 0.217)
NDE	−0.036	(−0.101, 0.030)
NIE	−0.116	(−0.154, −0.078)
PE(0)	0.720	(0.308, 1.132)
PE(1)	1.310	(0.087, 2.534)
PM	0.765	(0.365, 1.166)

TABLE 12.2
Mediation Analysis of the Double What-If? Study with Parametric Assumptions

Measure	Estimate	95% CI
TE	−0.151	(−0.202, −0.101)
CDE(0)	−0.039	(−0.099, 0.022)
CDE(1)	−0.039	(−0.100, 0.022)
NDE	−0.037	(−0.096, 0.021)
NIE	−0.114	(−0.144, −0.084)
PE(0)	0.744	(0.383, 1.106)
PE(1)	0.743	(0.379, 1.108)
PM	0.753	(0.403, 1.103)

We note that the results of the nonparametric and parametric mediation analyses are almost the same, except for the controlled direct effects, which are nearly the same and both negative in the parametric analysis due to the absence of an interaction between treatment and mediator in the outcome model.

We note that the PE(0), PE(1), and PM are all quite high, and correspondingly that the estimated direct effects are not statistically signifiant. Thus, the effect of naltrexone on unsuppressed viral load appears to be mediated almost entirely by reduced drinking. We know from the program that generated the data, `doublewhatifsim.r` in Chapter 1, that the effect is 100% mediated. Finally, we note that the confidence intervals for PE(0), PE(1), and PM include values greater than one, which in this case we should interpret as truncated at one.

For our second example, we simulated data to show that mediation analysis can lead to direct and indirect effects with opposite signs, which is perhaps nonintuitive. We used `simmed.r`, below.

```
simmed.r <- function ()
{
  set.seed(44444)
  A <- rbinom(n = 1000, size = 1, prob = .5)
  H <- rbinom(n = 1000, size = 1, prob = .5)
  # Let the mediator increase with A and/or H
  tmppm <- A + H
  pm <- exp(tmppm) / (1 + exp(tmppm))
  M <- rbinom(n = 1000, size = 1, prob = pm)
  # Let the outcome increase with A and/or H but decrease with M
  tmppy <- H - M + A
  py <- exp(tmppy) / (1 + exp(tmppy))
  Y <- rbinom(n = 1000, size = 1, prob = py)
  dat <- cbind(H, A, M, Y)
  dat <- data.frame(dat)
  dat
}
```

To motivate the simulation, we let A be randomization to a diet that does not work, in fact even causing weight gain for some participants. We let H indicate genetics inducing a propensity for weight gain. Let M be adoption of a stricter diet midway through the study. Let Y be weight at the end of study. We see that $A = 1$ and $H = 1$ are randomized with probability 0.5, and that they both influence $M = 1$. However, the end of study weight tends to be higher for those with a propensity for weight gain, with $H = 1$, and for those randomized to the diet, with $A = 1$, with some mitigation from adopting the stricter diet, $M = 1$.

The results of mediation analysis are presented in Table 12.3. We see that overall, the initial diet $A = 1$ causes weight gain, and that moreover, there is a direct effect of weight gain no matter which method we use to quantify the direct effect. However, the indirect effect is for weight loss via adoption of $M = 1$. The opposite signs of the direct and indirect effects lead to PE(0), PE(1), and PM that are not meaningful, and we see that their estimates are

TABLE 12.3
Nonparametric Mediation Analysis of the Data Generated by `simmed.r`

Measure	Estimate	95% CI
TE	0.160	(0.102, 0.218)
CDE(0)	0.279	(0.183, 0.376)
CDE(1)	0.168	(0.097, 0.239)
NDE	0.209	(0.153, 0.265)
NIE	−0.049	(−0.071, −0.028)
PE(0)	−0.747	(−1.478, −0.017)
PE(1)	−0.049	(−0.324, 0.225)
PM	−0.308	(−0.533, −0.083)

negative. Nevertheless, the mediation analysis is helpful in terms of teasing apart the direct and indirect effects of the initial diet $A = 1$.

12.4 Exercise

1. Use the complete case dataset from the GSS data analyzed in exercise 1 of Chapter 3, with variables `owngun`, `conservative`, `white`, `gt65`, and `female`, with the product method and the difference method to assessing whether the effect of `white` on `owngun` is mediated by `conservative`, adjusting for confounding by `gt65` and `female`.

2. Use parametric logistic models with the previous dataset to assess whether the effect of `white` on `owngun` is mediated by `conservative`, adjusting for confounding by `gt65` and `female`. Compute the TE, CDE(0), CDE(1), NDE, NIE, PE(0), PE(1), and PM, together with 95% confidence intervals.

3. Use the `brfss` data introduced in the exercises for Chapter 3 and analyzed again in the exercises for Chapter 6, and subset it to include only those with `gt65` equal to zero. Investigate whether the effect of `whitenh` on `flushot` is mediated by `insured`, adjusting for the confounders `female`, `gthsedu`, and `rural`, using the product method and the difference method.

4. Use parametric models with the previous dataset to estimate whether the effect of `whitenh` on `flushot` is mediated by `insured`, adjusting for confounding by `female`, `gthsedu`, and `rural`. Compute the TE, CDE(0), CDE(1), NDE, NIE, PE(0), PE(1), and PM, together with 95% confidence intervals.

5. Use `simmedex.r` to construct the dataset `medexdat`. Judging from the `R` code, is there any mediation by M of the effect of A on Y? Attempt to use the difference method to assess mediation, but with two parametric logistic models instead of with two parametric linear models. Is the difference in the two coefficients of A statistically significant, as assessed via the bootstrap? If so, this is due to *non-collapsibility* of the odds ratio. That is, the conditional odds ratio for the association of Y and A within each level of M can be constant and equal to o, but when we ignore M, or *collapse* over it, the unconditional odds ratio for the association of Y and A may not equal o. This can happen even when A is not associated with M, as in this example. This is why we cannot use the difference method with logistic regression models.

```
simmedex.r <- function ()
{
  set.seed(55)
  A <- rbinom(n = 1000, size = 1, prob = .5)
  M <- rbinom(n = 1000, size = 1, prob = .5)
  tmppy <- 1.39 - .985 * A - 1.795 * M
  py <- exp(tmppy) / (1 + exp(tmppy))
  Y <- rbinom(n = 1000, size = 1, prob = py)
  dat <- cbind(A, M, Y)
  dat <- data.frame(dat)
  dat
}
```

13

Adjusting for Time-Dependent Confounding

The methods presented in this chapter were motivated in Chapter 1 by the cancer clinical trial, with a hypothetical version of that data presented in Table 1.4. To analyze those data, we will assume the causal DAG in Figure 13.1 holds.

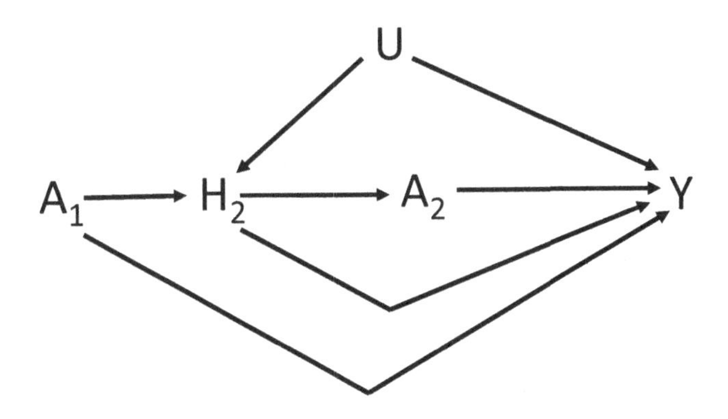

FIGURE 13.1: H_2 is a Time-Dependent Confounder

In the DAG, the variable H_2 is a *time-dependent confounder*, because it is a confounder for treatment A_2 at time two, but not for treatment A_1 at time one. In fact, H_2 is an *intermediate variable* for the effect of treatment A_1 on the outcome Y, because it is a mediator of that effect. This dual role of H_2 as a confounder with respect to one treatment but as an intermediate variable with respect to another treatment makes it tricky to adjust for it in the analysis of the joint effects of A_1 and A_2 on Y. In the following subsections, we will present three approaches to analyzing causal effects of A_1 and A_2 on Y. As we will see, the three approaches require three different potential outcomes frameworks for validity. In what follows, we will assume A_1 and A_2 are binary, noting that modifications exist for more general versions.

DOI: 10.1201/9781003146674-13

13.1 Marginal Structural Models

Marginal structural models (MSMs) (Robins et al. (2000)) require consistency of four potential outcomes $Y(a_1, a_2)$, for $a_1 = 0, 1$ and $a_2 = 0, 1$, to treatment with $A_1 = a_1$ followed by $A_2 = a_2$. We also assume sequential randomization,

$$A_1 \amalg Y(a_1, a_2) \tag{13.1}$$

and

$$A_2 \amalg Y(a_1, a_2)|A_1, H_2. \tag{13.2}$$

A marginal structural model is a model for the marginal means of the potential outcomes, such as the saturated model

$$E(Y(a_1, a_2)) = \beta_0 + \beta_1 a_1 + \beta_2 a_2 + \beta_3 a_1 a_2. \tag{13.3}$$

To estimate the parameter $\beta = (\beta_0, \beta_1, \beta_2, \beta_3)$, we need to relate $E(Y(a_1, a_2))$ to the observed data. One way to do that is through the relation

$$E(Y(a_1, a_2)) = E_{H_2|A_1=a_1} E(Y|A_1 = a_1, H_2, A_2 = a_2), \tag{13.4}$$

which we prove as follows. By (13.1), we have that

$$E(Y(a_1, a_2)) = E(Y(a_1, a_2)|A_1 = a_1).$$

Then by the double expectation theorem, we have that

$$E(Y(a_1, a_2)|A_1 = a_1) = E_{H_2|A_1=a_1} E(Y(a_1, a_2)|A_1 = a_1, H_2).$$

Finally, by (13.2), the right-hand side equals

$$E_{H_2|A_1=a_1} E(Y(a_1, a_2)|A_1 = a_1, H_2, A_2 = a_2),$$

which equals the right-hand side of (13.4) by consistency.

Thus, one way to estimate β is to use (13.4), which involves the outcome model $E(Y|A_1, H_2, A_2)$. However, a more popular method arises from an exposure-modeling approach, which involves a model for

$$e(H_2, A_1) = E(A_2|H_2, A_1).$$

We can use proofs directly analogous to those for (6.9) and (6.10) to show that

$$E(Y(a_1, 1)) = E\left(\frac{A_2 Y}{e(H_2, A_1)}|A_1 = a_1\right)$$

and

$$E(Y(a_1, 0)) = E\left(\frac{(1 - A_2)Y}{1 - e(H_2, A_1)}|A_1 = a_1\right).$$

We also use proofs directly analogous to those for (6.15) and (6.16) to show that weighted averages of Y with weights equal to

$$W = \frac{1}{A_2 e(H_2, A_1) + (1 - A_2)(1 - e(H_2, A_1))}$$

within the four $(A_1 = a_1, A_2 = a_2)$ subgroups are valid estimates of $E(Y(a_1, a_2))$. As in Chapter 6, we can compute these estimates using the `weighted.mean` function or using the `glm` or `gee` functions to fit a weighted linear model; the method is often called *inverse-probability of treatment weighted* (IPTW) estimation. We implement it with the `glm` function and the code `msm.r`, which outputs an estimate of β as well *contrasts* of β. For example, the contrast $\beta_1 + \beta_3$ compares administration of both $A_1 = 1$ and $A_2 = 1$ with administration of just $A_2 = 1$, whereas the contrast $\beta_2 + \beta_3$ compares $A_1 = 1$ and $A_2 = 1$ with just $A_1 = 1$. Finally, $\beta_1 + \beta_2 + \beta_3$ compares administration of both treatments with administration of neither treatment.

```
msm.r <- function(dat = cogdat,
                  ids = 1:nrow(cogdat))
{
  dat <- dat[ids, ]
  # Estimate the probability of treatment at time two
  e <- glm(A2 ~ A1 * H2, data = dat, family = binomial)$fitted
  # Estimate the weights
  W <- 1 / (e * dat$A2 + (1 - e) * (1 - dat$A2))
  # Fit the marginal structural model
  msm <- summary(glm(
    formula = Y ~ A1 + A2 + A1 * A2,
    data = dat,
    family = gaussian,
    weights = W
  ))
  beta <- msm$coef[, 1]
  # Return contrasts of the MSM parameters
  c(beta, (beta[2] + beta[4]), (beta[3] + beta[4]),
       (beta[2] + beta[3] + beta[4]))
}
```

We estimate the sampling distribution using the bootstrap. The hypothetical cancer clinical trial data are stored in `cogdat`. Results of applying marginal structural models with IPTW estimation to the cancer data are presented in Table 13.1.

We observe that the estimates of β_1 and β_2 are not statistically significantly different from zero. Therefore, either treatment on its own does not appear to influence survival at two years versus administration of neither treatment. However, the statistical significance of the estimates of $\beta_1 + \beta_3$ and $\beta_2 + \beta_3$ suggests that administering $A_1 = 1$ followed by $A_2 = 1$ results in

TABLE 13.1
IPTW of MSM parameters for the Hypothetical Cancer Clinical Trial

Parameter	Estimate	95% CI
β_0	0.261	(0.224, 0.298)
β_1	−0.008	(−0.063, 0.047)
β_2	−0.060	(−0.129, 0.009)
β_3	0.208	(0.056, 0.359)
$\beta_1 + \beta_3$	0.199	(0.056, 0.342)
$\beta_2 + \beta_3$	0.148	(0.008, 0.287)
$\beta_1 + \beta_2 + \beta_3$	0.140	(−0.001, 0.280)

increased survival relative to administering just one treatment. To compare joint administration to administration of neither treatment, we need to estimate the contrast $\beta_1+\beta_2+\beta_3$, which we see is not quite statistically significant, although it is very close. Our point estimates suggest that administering neither treatment results in a 26.1% chance of survival for two years, whereas administering both results in a 26.1%+14.0%=40% chance of survival for two years. These survival probabilities refer to the entire population that participants were sampled from; that is, they are marginal rather than conditional probabilities.

13.2 Structural Nested Mean Models

Structural nested mean models (SNMM) (Robins (1994)) require consistency of two potential outcomes: Y_1 is the potential outcome to treatment with the observed A_1 and then treatment with $A_2 = 0$, whereas Y_0 is the potential outcome to $A_1 = 0$ followed by $A_2 = 0$. We also assume sequential randomization,

$$A_1 \amalg Y_0 \tag{13.5}$$

and

$$A_2 \amalg Y_1 | A_1, H_2. \tag{13.6}$$

For our example, there are two levels to the 'nest' of the structural nested mean model. The second level reflects the effect of changing A_2 from its observed value to $A_2 = 0$:

$$E(Y - Y_1 | A_2, H_2, A_1) = f(A_1, H_2; \beta_2) A_2, \tag{13.7}$$

where $f(\cdot)$ is specified by the analyst, whereas the first level reflects the effect of further changing A_1 from its observed value to $A_1 = 0$:

$$E(Y_1 - Y_0 | A_1) = \beta_1 A_1. \tag{13.8}$$

We observe that these models target conditional causal effects, with the first one conditional on A_2, H_2, and A_1; the second one conditional on A_1. The parameters β_2 and β_1 quantify elements of those conditional causal effects.

To estimate β_1 and β_2, we need to relate them to the observed data. We can use the relations

$$E(Y|A_2, H_2, A_1) = f(A_1, H_2; \beta_2)A_2 + h(H_2, A_1) \tag{13.9}$$

and

$$E(Y - f(A_1, H_2; \beta_2)A_2|A_1) = \alpha + \beta_1 A_1, \tag{13.10}$$

where $h(\cdot)$ must be correctly specified by the analyst. For estimation of β_2 using the first relation, we can fit a non-parametric model if possible, or a parametric model if necessary. For estimation of β_1 in the second relation, we (a) substitute into the left-hand side the estimate $\hat{\beta}_2$ from the fitting the model specified by first relation, and (b) fit the model specified by the second relation to estimate β_1. To prove the first relation (13.9), it suffices to show that

$$E(Y - f(A_1, H_2; \beta_2)A_2|A_2, H_2, A_1) \tag{13.11}$$

does not depend on A_2. By (13.7), we have that (13.11) equals $E(Y_1|A_2, H_2, A_1)$, which, by (13.6), does not depend on $A2$.

To prove the second relation (13.10), it suffices to show that

$$E(Y - f(A_1, H_2; \beta_2)A_2 - \beta_1 A_1|A_1) \tag{13.12}$$

does not depend on A_1. By the double expectation theorem, we have that (13.12) equals

$$E_{A_2,H_2|A_1} E(Y - f(A_1, H_2)A_2|A_2, H_2, A_1) - \beta_1 A_1,$$

which, by (13.7), equals

$$E_{A_2,H_2|A_1} E(Y_1|A_2, H_2, A_1) - \beta_1 A_1 = E(Y_1|A_1) - \beta_1 A_1.$$

By (13.8), this equals $E(Y_0|A_1)$, which equals $E(Y_0)$ by (13.5), which does not depend on A_1.

We estimated the parameters of the structural nested mean model for the hypothetical cancer clinical trial using `snmm.r`, with bootstrap estimates of the sampling distribution. We fit the models

$$E(Y - Y_1|A_2, H_2, A_1) = (\beta_{20} + A_1\beta_{21} + H_2\beta 22 + A_1 * H_2\beta_{23})A_2,$$

and

$$E(Y_1 - Y_0|A_1) = \beta_1 A_1.$$

```
snmm.r <- function(dat = cogdat,
                   ids = 1:nrow(cogdat))
{
  dat <- dat[ids, ]
  # Fit the model for the second level of the nest
  nest2.out <- summary(lm(Y ~ A1 + H2 + A2 +
                A1 * H2 + A1 * A2 + H2 * A2 + A1 * H2 * A2,
                data = dat))
  b2 <- nest2.out$coef[, 1]
  # Estimate the potential outcome of setting A2 to zero
  Y1hat <-
  dat$Y - b2[4] * dat$A2 - b2[6] * dat$A1 * dat$A2 - b2[7] * dat$H2 *
   dat$A2 - b2[8] * dat$A1 * dat$H2 * dat$A2
  # Fit the model for the first level of the nest
  nest1.out <- summary(lm(Y1hat ~ A1, data = dat))
  # Estimate the effect of A1
  b1 <- nest1.out$coef[2, 1]
  # Return estimated effects
  c(b2[c(4, 6:8)], b1)
}
```

The parameter estimates and their 95% confidence intervals are presented in Table 13.2.

TABLE 13.2
SNMM Estimation for the Hypothetical Cancer Clinical Trial

Contrast	Estimate	95% CI
β_{20}	−0.293	(−0.337, −0.248)
$\beta_{20} + \beta_{22}$	0.479	(0.290, 0.669)
$\beta_{20} + \beta_{21}$	0.393	(0.159, 0.626)
$\beta_{20} + \beta_{21} + \beta_{22} + \beta_{23}$	−0.135	(−0.259, −0.011)
β_1	−0.008	(−0.063, 0.047)

From estimation of β_1, we see that the effect of A_1 is not statistically significant when A_2 will not be administered afterwards. From estimation of the contrasts of β_2, we see that the effect of A_2 is to decrease survival when $A_1 = 0$ and $H_2 = 0$ ($\hat{\beta}_{20} = -0.293$), to increase survival when $A_1 = 0$ and $H_2 = 1$ ($\hat{\beta}_{20} + \hat{\beta}_{22} = 0.479$), to increase survival when $A_1 = 1$ and $H_2 = 0$ ($\hat{\beta}_{20} + \hat{\beta}_{21} = 0.393$), and to decrease survival when $A_1 = 1$ and $H_2 = 1$ ($\hat{\beta}_{20} + \hat{\beta}_{21} + \hat{\beta}_{22} + \hat{\beta}_{23} = -0.135$). Therefore, the decision about whether to treat with $A_2 = 0$ or $A_2 = 1$ should make use of information on A_1 and H_2, where possible.

13.3 Optimal Dynamic Treatment Regimes

Marginal structural models provided information on optimal *static treatment regimes*, that is, the optimal sequence of A_1 and A_2 when a static decision must be made at baseline. Structural nested mean models provided some information on optimal *dynamic treatment regimes*; we learned how to choose the optimal A_2 when a dynamic decision could be made after observing A_1 and H_2. However, we were left hanging about the optimal choice of A_1. In this subsection, we learn how to estimate the *optimal dynamic treatment regime* (Murphy (2003); Moodie et al. (2007)), which will tell us how first to select A_1 and second to select A_2, taking into account A_1 and H_2, in order to obtain the best results for the population under study.

We require consistency of the eight potential outcomes $Y(a_1, d_2)$, where a_1 is a static treatment regime recording the setting of A_1 to 0 or 1, and d_2 is one of four possible dynamic treatment regimes recording the setting of A_2 to 0 or 1 depending on a_1 and H_2. For example, supposing $a_1 = 0$, then one of the four possible dynamic treatment regimes for d_2 is to set A_2 to 0 if $H_2 = 0$ or to 1 if $H_2 = 1$, whereas another possible one is to set A_2 to 0 regardless of H_2. We also assume sequential randomization,

$$A_1 \amalg Y(a_1, d_2) \tag{13.13}$$

and

$$A_2 \amalg Y(a_1, d_2) | A_1, H_2. \tag{13.14}$$

Given a_1, we determine the optimal d_2, which is a function of H_2, by maximizing

$$E(Y(a_1, d_2) | A_1 = a_1, H_2).$$

By (13.14), this is equivalent to maximizing

$$E(Y(a_1, d_2) | A_1 = a, H_2, A_2 = d_2),$$

which, by consistency, is equivalent to maximizing

$$E(Y | A_1 = a, H_2, A_2 = d_2), \tag{13.15}$$

which is a function of the observed data. In this way, we can determine the optimal d_2, which we denote by d_2^{opt}, as a function of a_1.

To find the optimal a_1, or a_1^{opt}, we find the a_1 that maximizes

$$E(Y(a_1, d_2^{opt})).$$

By (13.13), this is equivalent to maximizing

$$E(Y(a_1, d_2^{opt}) | A_1 = a_1),$$

which, by the double expectation theorem, equals

$$E_{H_2|A_1=a_1}E(Y(a_1,d_2^{opt})|A_1=a_1,H_2),$$

which, by (13.14), equals

$$E_{H_2|A_1=a_1}E(Y(a_1,d_2^{opt})|A_1=a_1,H_2,A_2=d_2^{opt}),$$

which, by consistency, is equivalent to maximizing

$$E_{H_2|A_1=a_1}E(Y|A_1=a_1,H_2,A_2=d_2^{opt}), \tag{13.16}$$

which is a function of the observed data.

Therefore, to find the optimal dynamic treatment regime (a_1^{opt}, d_2^{opt}), we first find the d_2 that maximizes (13.15) for each possible value of a_1, and we second find the a_1 that maximizes (13.16) with d_2^{opt} set equal to that d_2. We compute the optimal dynamic treatment regime using the series of R functions below. We start with the data set `cogdat`, in the form

```
> head(cogdat)
   A1 H2 A2 Y
1   0  0  0 1
11  0  0  0 1
12  0  0  0 1
13  0  0  0 1
14  0  0  0 1
15  0  0  0 1
```

Next, we use the function `mkcogtab.r` to convert the data set into table form:

```
mkcogtab.r <- function(dat = cogdat,
                       ids = 1:nrow(cogdat))
{
  dat <- dat[ids, ]
  tab <- data.frame(xtabs( ~ A2 + H2 + A1, data = dat))
  tmp <- data.frame(xtabs(Y ~ A2 + H2 + A1, data = dat))
  tab$prop <- tmp$Freq / tab$Freq
  tab
}
> mkcogtab.r()
  A2 H2 A1 Freq    prop
1  0  0  0  410 0.29268
2  1  0  0   30 0.00000
3  0  1  0  160 0.18750
4  1  1  0   30 0.66667
5  0  0  1  280 0.10714
6  1  0  1   20 0.50000
7  0  1  1  190 0.42105
8  1  1  1   70 0.28571
```

where prop is the proportion of participants in the A_2, H_2, A_1 stratum with $Y = 1$. Then, we find d_2^{opt} using the function A2opt.r:

```
A2opt.r <- function(tab)
{
  A2opt <- NULL
  propopt <- NULL
  # Calculate optimal A2 for A1=H2=0
  a <- which(tab[1:2, "prop"] == max(tab[1:2, "prop"])) - 1
  A2opt <- c(A2opt, a, a)
  # Calculate the conditional survival probability
  b <- tab$prop[(1 + a)]
  propopt <- c(propopt, b, b)
  # Calculate the optimal A2 for A1=0 H2=1
  a <- which(tab[3:4, "prop"] == max(tab[3:4, "prop"])) - 1
  A2opt <- c(A2opt, a, a)
  # Calculate the conditional survival probability
  b <- tab$prop[(3 + a)]
  propopt <- c(propopt, b, b)
  # Calculate the optimal A2 for A1=1 H2=0
  a <- which(tab[5:6, "prop"] == max(tab[5:6, "prop"])) - 1
  A2opt <- c(A2opt, a, a)
  # Calculate the conditional survival probability
  b <- tab$prop[(5 + a)]
  propopt <- c(propopt, b, b)
  # Calculate the optimal A2 for A1=H2=1
  a <- which(tab[7:8, "prop"] == max(tab[7:8, "prop"])) - 1
  A2opt <- c(A2opt, a, a)
  # Calculate the conditional survival probability
  b <- tab$prop[(7 + a)]
  propopt <- c(propopt, b, b)
  tab$A2opt <- A2opt
  tab$propA2opt <- propopt
  tab
}
> A2opt.r(mkcogtab.r())
  A2 H2 A1 Freq    prop A2opt propA2opt
1  0  0  0  410 0.29268     0   0.29268
2  1  0  0   30 0.00000     0   0.29268
3  0  1  0  160 0.18750     1   0.66667
4  1  1  0   30 0.66667     1   0.66667
5  0  0  1  280 0.10714     1   0.50000
6  1  0  1   20 0.50000     1   0.50000
7  0  1  1  190 0.42105     0   0.42105
8  1  1  1   70 0.28571     0   0.42105
```

where A2opt is d_2^{opt} as a function of A_1 and H_2. This enables us to find a_1^{opt} using A1opt.r:

```
A1opt.r <- function(tab)
```

```
{
  # Estimate E(H2|A1)
  a <- prop.table(xtabs(Freq ~ H2 + A1, data = tab), 2)
  # Find the optimal A1 and marginal survival probability
  A10prop <- a[1, 1] * tab$propA2opt[1] + a[2, 1] * tab$propA2opt[3]
  A11prop <- a[1, 2] * tab$propA2opt[5] + a[2, 2] * tab$propA2opt[7]
  if (A10prop > A11prop) {
    A1opta <- 0
    propA1opta <- A10prop
  }
  if (A11prop > A10prop) {
    A1opta <- 1
    propA1opta <- A11prop
  }
  A1opt <- rep(A1opta, 8)
  propA1opt <- rep(propA1opta, 8)
  tab$A1opt <- A1opt
  tab$propA1opt <- propA1opt
  tab
}
> A1opt.r(A2opt.r(mkcogtab.r()))
  A2 H2 A1 Freq      prop A2opt propA2opt A1opt propA1opt
1  0  0  0  410 0.29268     0   0.29268     1   0.46335
2  1  0  0   30 0.00000     0   0.29268     1   0.46335
3  0  1  0  160 0.18750     1   0.66667     1   0.46335
4  1  1  0   30 0.66667     1   0.66667     1   0.46335
5  0  0  1  280 0.10714     1   0.50000     1   0.46335
6  1  0  1   20 0.50000     1   0.50000     1   0.46335
7  0  1  1  190 0.42105     0   0.42105     1   0.46335
8  1  1  1   70 0.28571     0   0.42105     1   0.46335
```

where `A1opt` is a_1^{opt}, which is estimated as 1. Putting this together with the results of `A2opt`, we observe that the optimal dynamic treatment regime is to set $A_1 = 1$ and then to set $A_2 = 1$ if $H_2 = 0$ or $A_2 = 0$ if $H_2 = 1$. The marginal survival probability following implementation of the optimal dynamic treatment regime is estimated at 0.463. Conditional on H_2, this survival probability increases to 0.500 if $H_2 = 0$ and decreases to 0.421 if $H_2 = 1$.

Finally, we can use the bootstrap to determine the proportion of bootstrap samples with the bootstrap estimate of a_1^{opt} equal to the observed estimate of a_1^{opt}, and likewise we can compute bootstrap confidence intervals for the survival probability corresponding to the optimal treatments. Because we are using the bootstrap to compute estimates not returned by the `boot` function, we need to write our own bootstrap function using the `sample` command. The functions `optimal.r` and `bootoptimal.r` provide the estimates for us:

```
optimal.r <- function(tab)
{
  # Identify the optimal A1 and marginal survival probability
  A1opt <- tab$A1opt[1]
```

```
  propA1opt <- tab$propA1opt[1]
  # Identify the optimal dynamic regime and conditional survival
    probabilities
  #  if optimal A1 is 0
  if (A1opt == 0)
  {
    A2optH20 <- tab$A2opt[1]
    propA2optH20 <- tab$propA2opt[1]
    A2optH21 <- tab$A2opt[3]
    propA2optH21 <- tab$propA2opt[3]
  }
  # Identify the optimal dynamic regime and conditional survival
    probabilities
  #  if optimal A1 is 1
  if (A1opt == 1)
  {
    A2optH20 <- tab$A2opt[5]
    propA2optH20 <- tab$propA2opt[5]
    A2optH21 <- tab$A2opt[7]
    propA2optH21 <- tab$propA2opt[7]
  }
  c(A1opt,
    propA1opt,
    A2optH20,
    propA2optH20,
    A2optH21,
    propA2optH21)
}
bootoptimal.r <- function()
{
  # Find the estimates for the original sample
  orig <- optimal.r(A1opt.r(A2opt.r(mkcogtab.r())))
  dat <- cogdat
  Nboot <- 1000
  A1opt <- NULL
  propA1opt <- NULL
  propA2optH20 <- NULL
  propA2optH21 <- NULL
  for (i in 1:Nboot)
  {
    # Construct a bootstrap sample
    ids <- sample(c(1:nrow(cogdat)), replace = T)
    # Find the estimates for the bootstrap sample
    out <- optimal.r(A1opt.r(A2opt.r(mkcogtab.r(dat, ids))))
    A1opt <- c(A1opt, out[1])
    propA1opt <- c(propA1opt, out[2])
    propA2optH20 <- c(propA2optH20, out[4])
    propA2optH21 <- c(propA2optH21, out[6])
  }
```

```
  # Find the proportion of times optimal A1=1
  pA1opt <- mean(A1opt)
  # Estimate confidence intervals for marginal and
  #  conditional survival probabilities
  SEpropA1opt <- sqrt(var(propA1opt))
  lclpropA1opt <- orig[2] - 1.96 * SEpropA1opt
  uclpropA1opt <- orig[2] + 1.96 * SEpropA1opt
  SEpropA2optH20 <- sqrt(var(propA2optH20))
  lclpropA2optH20 <- orig[4] - 1.96 * SEpropA2optH20
  uclpropA2optH20 <- orig[4] + 1.96 * SEpropA2optH20
  SEpropA2optH21 <- sqrt(var(propA2optH21))
  lclpropA2optH21 <- orig[6] - 1.96 * SEpropA2optH21
  uclpropA2optH21 <- orig[6] + 1.96 * SEpropA2optH21
  list(
    pA1opt = pA1opt,
    lclpropA1opt = lclpropA1opt,
    uclpropA1opt = uclpropA1opt,
    lclpropA2optH20 = lclpropA2optH20,
    uclpropA2optH20 = uclpropA2optH20,
    lclpropA2optH21 = lclpropA2optH21,
    uclpropA2optH21 = uclpropA2optH21
  )
}
> bootoptimal.r()
$pA1opt
[1] 0.805
$lclpropA1opt
[1] 0.36094
$uclpropA1opt
[1] 0.56575
$lclpropA2optH20
[1] 0.25783
$uclpropA2optH20
[1] 0.74217
$lclpropA2optH21
[1] 0.18804
$uclpropA2optH21
[1] 0.65406
```

We observe that a_1^{opt} equals one for 80.5% of the bootstrap samples. We now can attach confidence intervals to our estimated survival probabilities. The marginal survival probability following implementation of the optimal dynamic treatment regime is estimated at 0.463 (0.361, 0.566). Conditional on H_2, this survival probability increases to 0.500 (0.258, 0.742) if $H_2 = 0$ and decreases to 0.421 (0.188, 0.654) if $H_2 = 1$. We observe that the survival probabilities conditional on H_2 are quite variable.

13.4 Exercises

1. To investigate the *sensitivity* of our analyses of the `cogdat` to setting $Y = 0$ for 30 patients with $A_1 = 0$, $H_2 = 0$, $A_2 = 1$, construct the dataset `scogdat` that sets $Y = 1$ instead for those 30. Repeat the marginal structural model analysis. Interpret your results.
2. Repeat the structural nested mean model analysis with `scogdat`. Interpret your results.
3. Determine the optimal dynamic treatment regime using `scogdat`. Interpret your results.
4. Create the dataset `simdat` using `sim.r`, where A_1 indicates randomization to consuming one tablespoon of apple cider vinegar in a cup of water one time per day from baseline to week 4, H_2 indicates 1% weight loss or more at week 4, A_2 indicates the choice to consume one tablespoon of apple cider vinegar from week 4 through week 7, and Y indicates 2% weight loss or more at week 7.

```
sim.r <- function ()
{
  set.seed(1313)
  U <- rbinom(n = 1000, size = 1, prob = .5)
  A1 <- rbinom(n = 1000, size = 1, prob = .5)
  H2prob <- A1 * .4 + U * .4
  H2 <- rbinom(n = 1000, size = 1, prob = H2prob)
  A2prob <- H2 * .6 + .2
  A2 <- rbinom(n = 1000, size = 1, prob = A2prob)
  Yprob <- A2 * .4 + A1 * .1 + U * .3
  Y <- rbinom(n = 1000, size = 1, prob = Yprob)
  dat <- cbind(A1, H2, A2, Y)
  dat <- data.frame(dat)
  dat
}
```

   Use the `R` code to compute $E(Y(a_1, a_2))$ for $a_1 = 0, 1$ and $a_2 = 0, 1$. Which static regime is best? Conduct a marginal structural model analysis to verify your response. Interpret your results.

5. Conduct a structural nested mean model analysis with `simdat`. Interpret your results.
6. Determine the optimal dynamic treatment regime using `simdat`. Interpret your results.

Appendix

You can download and install R at https://cran.r-project.org/, which also has links to manuals.

The R code in the book relies on the following extra packages, which must be installed and loaded previously: AER, boot, car, faraway, gee, geepack, Matching, resample. A package can be loaded like this:

```
> library(boot)
```

Bibliography

Abadie, A. (2005). Semiparametric Difference-in-Difference Estimators. *The Review of Economic Studies*, 72:1–19.

Anderson, D. (1964). Smoking and Respiratory Disease. *American Journal of Public Health*, 54:1856–1863.

Angrist, J., Imbens, G., and Rubin, D. (1996). Identification of Causal Effects Using Instrumental Variables. *Journal of the American Statistical Association*, 91:444–455.

Aristotle (350BC). *Physics, translated by R.P. Hardie and R.K. Gaye.*

Bang, H. and Robins, J. (2005). Doubly Robust Estimation in Missing Data and Causal Inference Models. *Biometrics*, 61:962–972.

Baron, R. and Kenny, D. (1986). The Moderator-Mediator Variable Distinction in Social Psychological Research: Conceptual, Strategic, and Statistical Considerations. *Journal of Personality and Social Psychology*, 51:1173–1182.

Beasley, C., Tollefson, G., and Tran, P. (1997). Safety of Olanzapine. *Journal of Clinical Psychiatry*, 58(suppl 10):13–17.

Bhatt, A. (2010). Evolution of Clinical Research: A History Before and Beyond James Lind. *Perspect Clin Res*, 1:6–10.

Blyth, C. (1972). On Simpson's Paradox and the Sure-Thing Principle. *Journal of the American Statistical Association*, 67:364–366.

Boden, J. and Fergusson, D. (2011). Alcohol and Depression. *Addiction*, 106:906–914.

Brown, S. (2005). *Scurvy: How a Surgeon, a Mariner, and a Gentleman Solved the Greatest Medical Mystery of the Age of Sail.* Thomas Dunne Books, New York, NY, USA.

Brumback, B. and Berg, A. (2008). On Effect-Measure Modification: Relationships Among Changes in the Relative Risk, Odds Ratio, and Risk Difference. *Statistics in Medicine*, 27:3453–3465.

Brumback, B., Bouldin, E., Zheng, H., Cannell, M., and Andresen, E. (2010). Testing and Estimating Model-Adjusted Effect-Measure Modification Using Marginal Structural Models and Complex Survey Data. *American Journal of Epidemiology*, 172:1085–1091.

Brumback, B., He, Z., Prasad, M., Freeman, M., and Rheingans, R. (2014). Using Structural Nested Models to Estimate the Effect of Cluster-level Adherence on Individual-level Outcomes with a Three-armed Cluster-randomized Trial. *Statistics in Medicine*, 33:1490–1502.

Brumback, B. and London, W. (2013). Causal Inference in Cancer Clinical Trials. In Tang, W. and Tu, W., editors, *Modern Clinical Trial Analysis*. Springer, New York, NY, USA.

Centers for Disease Control and Prevention (2020). Website: https://www.cdc.gov/brfss/annual_data/annual_2019.html.

Cheng, P. (1997). From Covariation to Causation: A Causal Power Theory. *Psychological Review*, 104:367–405.

Cole, S. and Frangakis, C. (2009). The Consistency Statement in Causal Inference: A Definition or an Assumption? *Epidemiology*, 20:3–5.

Cook, R. and Sackett, D. (1995). The Number Needed to Treat: A Clinically Useful Measure of Treatment Effect. *BMJ*, 310:452.

Cook, R., Zhou, Z., Miguez, M., Quiros, C., Espinoza, L., Lewis, J., Brumback, B., and Bryant, K. (2019). Reduction in Drinking was Associated with Improved Clinical Outcomes in Women with HIV Infection and Unhealthy Alcohol Use: Results From a Randomized Clinical Trial of Oral Naltrexone Versus Placebo. *Alcoholism: Clinical and Experimental Research*, 43:1790–1800.

Efron, B. and Stein, C. (1981). The Jackknife Estimate of Variance. *Annals of Statistics*, 9:586–596.

Efron, B. and Tibshirani, R. (1993). *An Introduction to the Bootstrap*. Chapman & Hall, New York, NY, USA.

Faraway, J. (2016). *Extending the Linear Model with R, 2nd Edition*. Chapman & Hall, Boca Raton, FL, USA.

Francois, B., Jeannet, R., Daix, T., Walton, A., Shotwell, M., Unsinger, J., Monneret, G., Rimmele, T., Blood, T., Morre, M., Gregoire, A., Mayo, G., Blood, J., Durum, S., Sherwood, E., and Hotchkiss, R. (2018). Interleukin-7 restores lymphocytes in septic shock: the IRIS-7 randomized clinical trial. *JCI Insight*, 3:1–18.

Frangakis, C. and Rubin, D. (2002). Principal Stratification in Causal Inference. *Biometrics*, 58:21–29.

Glynn, A. and Kashin, K. (2018). Front-Door Versus Back-Door Adjustment With Unmeasured Confounding: Bias Formulas for Front-Door and Hybrid Adjustments With Application to a Job Training Program. *Journal of the American Statistical Association*, 113:1040–1049.

Greenland, S., Pearl, J., and Robins, J. (1999). Causal Diagrams for Epidemiologic Research. *Epidemiology*, 10:37–48.

Greenland, S. and Robins, J. (2009). Identifiability, Exchangeability and Confounding Revisited. *Epidemiologic Perspectives & Innovations*, 6:1–9.

Hansen, B. (2008). The Prognostic Analogue of the Propensity Score. *Biometrika*, 95:481–488.

Helian, S., Brumback, B., Freeman, M., and Rheingans, R. (2016). Structural Nested Models for Cluster-Randomized Trials. In He, H., Wu, P., and Chen, D., editors, *Statistical Causal Inferences and Their Applications in Public Health Research*. Springer International Publishing, Switzerland.

Hernan, M. and Robins, J. (2020). *Causal Inference: What If*. Chapman and Hall, Boca Raton, FL, USA.

Hill, A. (1965). The Environment and Disease: Association or Causation? *Proceedings of the Royal Society of Medicine*, pages 7–12.

Hill, A. (1990). Suspended Judgement: Memories of the British Streptomycin Trial in Tuberculosis, The First Randomized Clinical Trial. *Controlled Clinical Trials*, 11:77–79.

Holland, P. (1986). Statistics and Causal Inference. *Journal of the American Statistical Association*, 81:945–960.

Hume, D. (1738). *A Treatise of Human Nature*.

Imbens, G. and Rubin, D. (1997). Bayesian Inference for Causal Effects in Randomized Experiments with Noncompliance. *The Annals of Statistics*, 25:305–327.

Imbens, G. and Rubin, D. (2015). *Causal Inference for Statistics, Social, and Biomedical Sciences: An Introduction*. Cambridge University Press, New York, NY, USA.

Jiang, A. and VanderWeele, T. (2015). When is the Difference Method Conservative for Assessing Mediation. *American Journal of Epidemiology*, 182:105–108.

Jo, B. and Stuart, E. (2009). On the Use of Propensity Scores in Principal Causal-Effect Estimation. *Statistics in Medicine*, 28:2857–2875.

Kohen, D. (2004). Diabetes Mellitus and Schizophrenia: Historical Perspective. *British Journal of Psychiatry*, 184(suppl 47):s64–s66.

Lambert, B., Cunningham, F., Miller, D., Dalack, G., and Hur, K. (2006). Diabetes Risk Associated with Olanazapine, Quetiapine, and Risperidone in Veterans Health Administration Patients with Schizophrenia. *American Journal of Epidemiology*, 164:672–681.

Lee, I. and Skerrett, P. (2001). Physical Activity and All-Cause Mortality: What is the Dose-Response Relation? *Medicine & Science in Sports & Exercise*, 33:S459–S471.

Lehmann, E. (1999). *Elements of Large-Sample Theory.* Springer, New York, NY, USA.

Loftus, T., Mira, J., Ozrazgat-Baslanti, T., Ghita, G., Wang, Z., Stortz, J., Brumback, B., Bihorac, A., Segal, M., Anton, S., Leeuwenburgh, C., Mohr, A., Efron, P., Moldawer, L., Moore, F., and Brakenridge, S. (2017). Sepsis and Critical Illness Research Center Investigators: Protocols and Standard Operating Procedures for a Prospective Cohort Study of Sepsis in Critically Ill Surgical Patients. *BMJ Open*, 7:1–8.

London, W., Frantz, C., Campbell, L., Seeger, R., Brumback, B., Cohn, S., Matthay, K., Castleberry, R., and Diller, L. (2010). Phase II Randomized Comparison of Topotecan Plus Cyclophosphamide Versus Topotecan Alone in Children with Recurrent or Refractory Neuroblastoma: A Children's Oncology Group Study. *Journal of Clinical Oncology*, 28:3808–3815.

Mill, J. (1846). *A System of Logic, Ratiocinative and Inductive, Being a Connected View of the Principles of Evidence and the Methods of Scientific Investigation.* Harper & Brothers, New York, NY, USA.

Mills, J. and Birks, M., editors (2014). *Qualitative Methodology: A Practical Guide.* Sage, Los Angeles, CA, USA.

Molyneux, P., Reghezza, A., and Xie, R. (2019). Bank Margins and Profits in a World of Negative Rates. *Journal of Banking and Finance*, 107:1–20.

Moodie, E., Richardson, T., and Stephens, D. (2007). Demystifying Optimal Dynamic Treatment Regimes. *Biometrics*, 63:447–455.

Morgan, S. and Winship, C. (2015). *Counterfactuals and Causal Inference: Methods and Principles for Social Research, 2nd Edition.* Cambridge University Press, New York, NY, USA.

Murphy, S. (2003). Optimal Dynamic Treatment Regimes. *Journal of the Royal Statistical Society, Series B*, 65:331–355.

National Institute on Alcohol Abuse and Alcoholism (2016). Helping Patients Who Drink Too Much: A Clinician's Guide.

Neyman, J., Dabrowska, D., and Speed, T. (1990). On the Application of Probability Theory to Agricultural Experiments. Essay on Principles. Section 9. Reprinted or adapted with permission of the Institute of Mathematical Statistics. *Statistical Science*, 5:465–472.

Pearl, J. (1995). Causal Diagrams for Empirical Research. *Biometrika*, 82:669–688.

Pearl, J. (1999). Probabilities of Causation: Three Counterfactual Interpretations and Their Identification. *Synthese*, 121:93–149.

Pearl, J. (2009). *Causality: Models, Reasoning, Inference, 2nd Edition.* Cambridge University Press, New York, NY, USA.

Pearl, J., Glymour, M., and Jewell, N. (2016). *Causal Inference in Statistics – A Primer.* Wiley, Chichester, West Sussex, U.K.

Pearl, J. and Mackenzie, D. (2018). *The Book of Why: The New Science of Cause and Effect.* Basic Books, New York, NY, USA.

Peduzzi, P., Concato, J., Kemper, E., Holford, T., and Feinstein, A. (1996). A Simulation Study of the Number of Events per Variable in Logistic Regression Analysis. *Journal of Clinical Epidemiology*, 49:1373–1379.

RECOVERY Collaborative Group (2020). Effect of Dexamethasone in Hospitalized Patients with COVID-19 – Preliminary Report.

Robins, J. (1989). The Control of Confounding by Intermediate Variables. *Statistics in Medicine*, 8:679–701.

Robins, J. (1994). Correcting for Non-Compliance in Randomized Trials Using Structural Nested Mean Models. *Communications in Statistics – Theory and Methods*, 23:2379–2412.

Robins, J., Hernan, M., and Brumback, B. (2000). Marginal Structural Models and Causal Inference in Epidemiology. *Epidemiology*, 11:550–560.

Rosenbaum, P. (2017). *Observation and Experiment.* Harvard University Press, Cambridge, MA, USA.

Rosenbaum, P. and Rubin, D. (1983). The Central Role of the Propensity Score in Observational Studies for Causal Effects. *Biometrika*, 70:41–55.

Rothman, K. and Greenland, S. (1998). *Modern Epidemiology, 2nd Edition.* Lippincott-Raven, Philadelphia, PA, USA.

Rothman, K., Greenland, S., and Lash, T. (2008). *Modern Epidemiology, 3rd Edition.* Lippincott Williams & Wilkins, Philadelphia, PA, USA.

Rubin, D. (1990). Comment. *Statistical Science*, 5:472–480.

Scanlan, J. (2006). Measuring Health Disparities. *Journal of Public Health Management & Practice*, 12:296.

Sebat, J., Lakshmi, B., Malhotra, D., Troge, J., Lese-Martin, C., Walsh, T., and Yamrom, B. (2007). Strong Associations of De Novo Copy Number Mutations with Autism. *Science*, 316:445–449.

Shannin, J. and Brumback, B. (2021). Disagreement Concerning Effect-Measure Modification. Website: https://arxiv.org/abs/2105.07285v1.

Slone, D., Shapiro, S., Miettinen, O., Finkle, W., and Stolley, P. (1979). Drug Evaluation After Marketing. *Annals of Internal Medicine*, 90:247–261.

Sofer, T., Richardson, D., Colicino, E., Schwartz, J., and Tchetgen Tchetgen, E. (2016). On Negative Outcome Control of Unobserved Confounding as a Generalization of Difference-in-Differences. *Statistical Science*, 31:348–361.

Sommer, A. and Zeger, S. (1991). On Estimating Efficacy from Clinical Trials. *Statistics in Medicine*, 10:45–52.

Stat News (2020). Website: https://www.statnews.com/2020/11/18/pfizer-biontech-covid19-vaccine-fda-data/.

The Alpha-Tocopherol Beta Carotene Cancer Prevention Study Group (1994). The Effect of Vitamin E and Beta Carotene on the Incidence of Lung Cancer and Other Cancers in Male Smokers. *The New England Journal of Medicine*, 330:1029–1035.

United Nations (2020). Website: https://population.un.org/wpp/DataQuery.

United Nations, Department of Economic and Social Affairs, Population Division (2019). World Mortality 2019: Data Booklet.

van der Vaart, A. (1998). *Asymptotic Statistics.* Cambridge University Press, Cambridge, UK.

Vanderweele, T. (2015). *Explanation in Causal Inference: Methods for Mediation and Interaction.* Oxford University Press, Oxford, U.K.

VanderWeele, T. and Robins, J. (2007). The Identification of Synergism in the Sufficient-Component-Cause Framework. *Epidemiology*, 18:329–339.

VanderWeele, T. and Vansteelandt, S. (2009). Conceptual Issues Concerning Mediation, Interventions, and Composition. *Statistics and its Interface*, 2:457–468.

Vansteelandt, S. and Goetghebeur, E. (2003). Causal Inference with Generalized Structural Mean Models. *Journal of the Royal Statistical Society, Series B*, 65:817–835.

Vittinghoff, E. and McCulloch, C. (2006). Relaxing the Rule of Ten Events per Variable in Logistic and Cox Regression. *American Journal of Epidemiology*, 165:710–718.

Waters, D., Alderman, E.L.and Hsia, J., Howard, B., Cobb, F., Rogers, W., Ouyang, P., Thompson, P., Tardif, J., Higginson, L., Bittner, V., Steffes, M., Gordon, D., Proschan, M., Younes, N., and Verter, J. (2002). Effects of Hormone Replacement Therapy and Antioxidant Vitamin Supplements on Coronary Atherosclerosis in Postmenopausal Women: A Randomized Controlled Trial. *Journal of the American Medical Association*, 288:2432–2440.

Wright, S. (1934). The Method of Path Coefficients. *The Annals of Mathematical Statistics*, 5:161–215.

Zeger, S., Liang, K., and Albert, P. (1988). Models for Longitudinal Data: A Generalized Estimating Equation Approach. *Biometrics*, 44:1049–1060.

Zhang, J. and Spirtes, P. (2008). Detection of Unfaithfulness and Robust Causal Inference. *Minds & Machines*, 18:239–271.

Zheng, H., Brumback, B., Lu, X., Bouldin, E., Cannell, M., and Andresen, E. (2013). Doubly Robust Testing and Estimation of Model-Adjusted Effect-Measure Modification with Complex Survey Data. *Statistics in Medicine*, 32:673–684.

Index

Printed in the United States
by Baker & Taylor Publisher Services